ASCE STANDARD

ASCE/EWRI 12-05
ASCE/EWRI 13-05
ASCE/EWRI 14-05

American Society of Civil Engineers

Standard Guidelines for the Design of Urban Subsurface Drainage
ASCE/EWRI 12-05

Standard Guidelines for the Installation of Urban Subsurface Drainage
ASCE/EWRI 13-05

Standard Guidelines for the Operation and Maintenance of Urban Subsurface Drainage
ASCE/EWRI 14-05

This document uses both the International System of Units (SI) and customary units.

Urban Drainage Standards Committee of the
Standards Development Council of the
Environmental and Water Resources Institute of the
American Society of Civil Engineers

Published by the American Society of Civil Engineers

Library of Congress Cataloging-in-Publication Data

ASCE/EWRI 12-05, ASCE/EWRI 13-05, ASCE/EWRI 14-05 : standard guidelines
for the design of urban subsurface drainage, ASCE/EWRI 12-05 ; standard
guidelines for the installation of urban subsurface drainage, ASCE/EWRI
13-05 ; standard guidelines for the operation and maintenance of urban
subsurface drainage, ASCE/EWRI 14-05 / Urban Drainage Standards
Committee of the Codes and Standards Activities Committee of the
Environmental and Water   Resources Institute of the American Society of
Civil Engineers.
     p. cm.
 Includes bibliographical references and index.
 ISBN 0-7844-0811-4
 1. Subsurface drainage—Standards—United States. 2.
Sewerage—Standards—United States. I. Title: Standard guidelines for
the design of urban subsurface drainage, ASCE/EWRI 12-05. II. Title:
Standard guidelines for the installation of urban subsurface drainage,
ASCE/EWRI 13-05. III. Title: Standard guidelines for the operation and
maintenance of urban subsurface drainage, ASCE/EWRI 14-05. IV. American
Society of Civil Engineers. V. American Society of Civil Engineers.
Urban Drainage Standards Committee. VI. Environmental and Water
Resources Institute (U.S.)

  TC970.A83 2005
  628'.20917320218—dc22

                                                      2005027802

Published by American Society of Civil Engineers
1801 Alexander Bell Drive
Reston, Virginia 20191
www.pubs.asce.org

Any statements expressed in these materials are those of the individual authors and
do not necessarily represent the views of ASCE, which takes no responsibility for
any statement made herein. No reference made in this publication to any specific
method, product, process, or service constitutes or implies an endorsement, recom-
mendation, or warranty thereof by ASCE.

ASCE makes no representation or warranty of any kind, whether express or implied,
concerning the accuracy, completeness, suitability, or utility of any information,
apparatus, product, or process discussed in this publication, and assumes no liability
therefore. This information should not be used without first securing competent
advice with respect to its suitability for any general or specific application. Anyone
utilizing this information assumes all liability arising from such use, including but
not limited to infringement of any patent or patents.

ASCE and American Society of Civil Engineers—Registered in U.S. Patent and
Trademark Office.

Photocopies: Authorization to photocopy material for internal or personal use under
circumstances not falling within the fair use provisions of the Copyright Act is
granted by ASCE to libraries and other users registered with the Copyright
Clearance Center (CCC) Transactional Reporting Service, provided that the base fee
of $25.00 per article is paid directly to CCC, 222 Rosewood Drive, Danvers, MA
01923. The identification for this book is 07844-0811-4/06/ $25.00. Requests for
special permission or bulk copying should be addressed to Permissions & Copyright
Dept., ASCE.

STANDARDS

In April 1980, the Board of Direction approved ASCE Rules for Standards Committees to govern the writing and maintenance of standards developed by the Society. All such standards are developed by a consensus standards process managed by the Codes and Standards Activities Committee. The consensus process includes balloting by the Balanced Standards Committee, which is composed of Society members and nonmembers, balloting by the membership of ASCE as a whole, and balloting by the public. All standards are updated or reaffirmed by the same process at intervals not exceeding 5 years.

The following Standards have been issued:

ANSI/ASCE 1-82 N-725 Guideline for Design and Analysis of Nuclear Safety Related Earth Structures

ANSI/ASCE 2-91 Measurement of Oxygen Transfer in Clean Water

ANSI/ASCE 3-91 Standard for the Structural Design of Composite Slabs and ANSI/ASCE 9-91 Standard Practice for the Construction and Inspection of Composite Slabs

ASCE 4-98 Seismic Analysis of Safety-Related Nuclear Structures

Building Code Requirements for Masonry Structures (ACI 530-02/ASCE 5-02/TMS 402-02) and Specifications for Masonry Structures (ACI 530.1-02/ASCE 6-02/TMS 602-02)

ASCE/SEI 7-05 Minimum Design Loads for Buildings and Other Structures

ANSI/ASCE 8-90 Standard Specification for the Design of Cold-Formed Stainless Steel Structural Members

ANSI/ASCE 9-91 listed with ASCE 3-91

ASCE 10-97 Design of Latticed Steel Transmission Structures

SEI/ASCE 11-99 Guideline for Structural Condition Assessment of Existing Buildings

ASCE 12-05 Guideline for the Design of Urban Subsurface Drainage

ASCE 13-05 Standard Guidelines for Installation of Urban Subsurface Drainage

ASCE 14-05 Standard Guidelines for Operation and Maintenance of Urban Subsurface Drainage

ASCE 15-98 Standard Practice for Direct Design of Buried Precast Concrete Pipe Using Standard Installations (SIDD)

ASCE 16-95 Standard for Load Resistance Factor Design (LRFD) of Engineered Wood Construction

ASCE 17-96 Air-Supported Structures

ASCE 18-96 Standard Guidelines for In-Process Oxygen Transfer Testing

ASCE 19-96 Structural Applications of Steel Cables for Buildings

ASCE 20-96 Standard Guidelines for the Design and Installation of Pile Foundations

ASCE 21-96 Automated People Mover Standards—Part 1

ASCE 21-98 Automated People Mover Standards—Part 2

ASCE 21-00 Automated People Mover Standards—Part 3

SEI/ASCE 23-97 Specification for Structural Steel Beams with Web Openings

ASCE/SEI 24-05 Flood Resistant Design and Construction

ASCE 25-97 Earthquake-Actuated Automatic Gas Shut-Off Devices

ASCE 26-97 Standard Practice for Design of Buried Precast Concrete Box Sections

ASCE 27-00 Standard Practice for Direct Design of Precast Concrete Pipe for Jacking in Trenchless Construction

ASCE 28-00 Standard Practice for Direct Design of Precast Concrete Box Sections for Jacking in Trenchless Construction

SEI/ASCE/SFPE 29-99 Standard Calculation Methods for Structural Fire Protection

SEI/ASCE 30-00 Guideline for Condition Assessment of the Building Envelope

SEI/ASCE 31-03 Seismic Evaluation of Existing Buildings

SEI/ASCE 32-01 Design and Construction of Frost-Protected Shallow Foundations

EWRI/ASCE 33-01 Comprehensive Transboundary International Water Quality Management Agreement

EWRI/ASCE 34-01 Standard Guidelines for Artificial Recharge of Ground Water

EWRI/ASCE 35-01 Guidelines for Quality Assurance of Installed Fine-Pore Aeration Equipment

CI/ASCE 36-01 Standard Construction Guidelines for Microtunneling

SEI/ASCE 37-02 Design Loads on Structures During Construction

CI/ASCE 38-02 Standard Guideline for the Collection and Depiction of Existing Subsurface Utility Data

EWRI/ASCE 39-03 Standard Practice for the Design and Operation of Hail Suppression Projects

ASCE/EWRI 40-03 Regulated Riparian Model Water Code

ASCE/EWRI 42-04 Standard Practice for the Design and Operation of Precipitation Enhancement Projects

ASCE/SEI 43-05 Seismic Design Criteria for Structures, Systems, and Components in Nuclear Facilities

ASCE/EWRI 44-05 Standard Practice for the Design and Operation of Supercooled Fog Dispersal Projects

Standard Guidelines for the Design of Urban Subsurface Drainage

CONTENTS

FOREWORD

The *Standard Guidelines for the Design of Urban Subsurface Drainage* is an independent document intended to complement the ASCE Manuals and Reports on Engineering Practice No. 95, *Urban Subsurface Drainage*. These standard guidelines are companions to the *Standard Guidelines for the Installation of Urban Subsurface Drainage* and *Standard Guidelines for the Operation and Maintenance of Urban Subsurface Drainage*. These standard guidelines were developed by the Urban Drainage Standards Committee, which is responsible to the Environmental and Water Resources Institute of the American Society of Civil Engineers.

The material presented in this publication has been prepared in accordance with recognized engineering principles. These standard guidelines should be used only under the direction of an engineer who is competent in the field of urban subsurface drainage. The publication of the material contained herein is not intended as a representation or warranty on the part of the American Society of Civil Engineers, or of any other person named herein, that this information is suitable for any general or particular use, or promises freedom from infringement of any patent or patents. Anyone making use of this information assumes all liability from such use.

ACKNOWLEDGEMENTS

The American Society of Civil Engineers (ASCE) acknowledges the work of the Urban Drainage Standards Committee of the Environmental and Water Resources Institute of ASCE (EWRI of ASCE).

This group comprises individuals from many backgrounds, including consulting engineering, research, the construction industry, education, and government. Those individuals who serve on the Urban Drainage Standards Committee are:

William Curtis Archdeacon, Chair
Richard H. Berich
Christopher B. Burke
Robert T. Chuck
F. Scott Dull
Robert S. Giurato, Secretary
S. David Graber
Jay M. Herskowitz
Conrad G. Keyes, Jr.
John M. Kurdziel
John J. Meyer
Philip M. Meyer
James R. Noll

Walter J. Ochs
Garvin J. Pederson
Glen D. Sanders
Erez Sela
Alan N. Sirkin
Edward L. Tharp
William J. Weaver
Richard D. Wenberg, Past Chair
David L. Westerling
Stan E. Wildesen
Lyman S. Willardson (deceased)
Donald E. Woodward

Standard Guidelines for the Design of Urban Subsurface Drainage

1.0 SCOPE

The intent of this standard is to present state-of-the-art design guidance for urban subsurface drainage in a logical order. The collection and conveyance of subsurface drainage waters are within the purview of this standard for applications such as airports, roads, and other transportation systems, as well as industrial, commercial, residential, and recreational areas. Incidental surface water is considered.

This standard does not address agricultural drainage, landfills, recharge systems, detention ponds, conventional storm sewer design, or the use of injection systems.

Customary units and standard international (SI) units are used throughout this document.

2.0 DEFINITIONS

2.1 GENERAL

This section defines specific terms for use in this standard. The references listed in Section 10 may be useful to augment understanding the terms in this standard.

2.2 TERMS

AOS—Apparent opening size of geotextiles, sometimes referred to as EOS (effective opening size).

Aquifer—Water-bearing stratum of permeable rock, sand, or gravel.

Barrier layer—Stratum having a hydraulic conductivity less than 10% of the weighted permeability of all the strata above it. Also referred to as the relative barrier.

Base drainage system—Permeable drainage blanket under a roadway pavement system.

Bedding—Granular material placed around subsurface drains to provide a structural support for the drain.

Chimney drain—Subsurface interceptor drain frequently used in dams, embankments, and similar construction to control seepage within the earthen structure. Chimney drains are constructed in near-vertical orientation and discharge to outlets at lower elevations.

Colloidal fines—Clay particles smaller than two microns.

Consolidation drains—Wick drains that allow hydrostatic pressure to be relieved, thus allowing material to consolidate.

Drain envelope—Generic name for materials placed on or around a drainage product, irrespective of whether used for mechanical support, hydraulic purposes (hydraulic envelope), or to stabilize surrounding soil material (filter envelope). Natural granular materials can also be used to improve bedding and backfill conditions.

Drainable water—Water that readily drains from soil under the influence of gravity. Also known as drainable porosity.

Evapotranspiration—Combined process of soil moisture evaporation and transpiration from plants.

Filter envelope—Permeable material placed around a drainage product to stabilize the structure of the surrounding soil material. A filter envelope may initially allow some fines and colloidal material to pass through it and into the drain.

Frost action—Movement of soil caused by freezing and thawing of soil moisture.

Geocomposite—Geosynthetic materials for collecting and transporting water while maintaining soil stability.

Geology—Natural subsurface soil and rock formations.

Geomembrane—Sheet material intended to form an impervious barrier.

Geosynthetic—Synthetic material or structure used as an integral part of a project, structure, or system. Within this category are subsurface drainage and water control materials such as geomembranes, geotextiles, and geocomposites.

Geotextile—Woven or nonwoven engineered fabric intended to allow the passage of water and limited soil particles.

High column edge drains—Composite drain systems in which the height of the drain system is greater than the width.

Hydraulic barrier layer—Soil stratum with a permeability less than 10% of the soil weighted permeability of the strata above it.

Hydraulic conductivity—See Permeability.

Hydraulic envelope—Permeable material placed around a drainage product to improve flow conditions in the area immediately adjacent to the drain.

Hydrology—Study of the movement of water in nature.

Infiltration—Passage of water into the soil.

Longitudinal drainage system—Drainage system parallel to a pavement system.

Perched water table—Localized condition of free water held in a pervious stratum because of an underlying impervious stratum.

Percolation—Downward movement of water through soil due to gravity.

Permeability (coefficient of permeability)—Rate at which a fluid moves through porous media under unit hydraulic gradient. In this publication, the fluid is always water and the media are always soil or rock.

Permittivity—Measure of the ability of a geo-textile to permit water flow perpendicular to its plane.

Phreatic surface—Upper surface of an unconfined body of groundwater.

Relief drain—Any product or construction that accelerates the removal of drainable subsurface water.

Seepage—Movement of drainable water through soil and rock.

Sink—Surface depression that allows surface drainage to enter the subsurface water system.

Soil texture—Relative proportions of sand, silt, and clay particles in a soil mass.

Subsurface water—All water beneath the ground or pavement surface. Sometimes referred to as groundwater.

Transverse drainage system—Drainage system usually at some angle to a roadway.

Water table—Upper limit of free water in a saturated soil or underlying material.

3.0 SITE ANALYSIS

3.1 GENERAL

Site analysis involves a thorough review of existing information on the site and its surrounding area. Additional studies normally are needed to fill voids in data required to prepare proper design and contract documents.

3.2 BASIC REQUIREMENTS

3.2.1 General
The basic information necessary for the design of all subsurface drainage facilities requires investigations of the following areas: topography, geography, water table, geology, water source, soil characteristics, environmental factors, and physical constraints.

3.2.2 Topography
All features that could influence subsurface drain location, installation, or operation must be considered in the design phase. Topographic studies of the drainage area ascertain the runoff direction following rainfalls, with lower elevations having greater drainage needs. A general topographic map of the project site and its surrounding area is required at the preliminary design stage. Topographic features of the area are essential in determining water movement to and from the drainage area. Surface water concentrations frequently result from watershed regions contributing to the area of concern.

A detailed topographic survey locating both planimetric features such as trees, ponds, ditches, culverts and catch basins, buildings, roads, walks, overhead utilities, and surface components of underground utilities, and establishing elevations of such, is necessary to develop and complete the design. The topographic information is used to establish proposed grades for the drain lines and outlet location, as well as to determine the necessity and locations of pump stations.

Additional research may be required of record information to avoid conflicts not apparent by physical evidence.

3.2.3 Geography
Geographic considerations may influence subsurface drainage system design. Coastal areas, floodplains, uplands, glaciated areas, humid areas, arid areas, and many other geographically distinct areas have unique drainage characteristics that the designer must take into account during the design process. The designer is expected to become familiar with these unique characteristics before proceeding with the design of drainage facilities.

3.2.4 Water Table
The site may require an evaluation of the underlying water table early in the design stage. The designer must know the type of water table (confined or unconfined), the depth of the water table below the surface, the direction and gradient of groundwater flow, and any seasonal fluctuations in elevation.

Information related to water table fluctuation throughout the year must be evaluated. An understanding of lag time for water table response after precipitation events and fluctuations related to well pumping in the vicinity of the proposed subsurface drainage site may be necessary. Water table characteristics are particularly essential when the surface of a water table is above or in proximity to the anticipated subsurface drain outlet. It is important that the receiving drain has adequate capacity to accept this subsurface discharge continuously, particularly through periods of wet weather. Hydrostatic heads from confined aquifers, perched water tables, heterogeneous soil, and disturbed soil are important and should be noted in the site analysis reports. Where flat gradients exist, the direction of flow can be influenced by the orientation of subsurface drains, but where steep gradients exist, the subsurface drains must be oriented so as to intercept the flow. Drains oriented perpendicular to natural flow are the most efficient. Efficiency drops off rapidly as the orientation of the drains becomes parallel to the direction of flow.

With this information, the designer can evaluate the impact the subsurface drain will have on the water table. The design of the drain must be adequate to maintain the water table below desired limits during the design recharge event. The designer may perform a risk assessment, which predicts the frequency with which the water table can be expected to rise above the desired limits.

3.2.5 Geology

Geology plays an important role in slope stability analysis where subsurface drains often are needed to stabilize slopes. Relative permeability and direction of flow in earth materials that may affect the behavior of subsurface drainage are of critical importance in subsurface drain design.

Geologic features such as rock layers, tight soil barriers, seepage paths, trench wall stability, and potentials for subsidence and collapse must be investigated.

3.2.6 Water Source

Present and projected water sources into the subsurface drain include precipitation; irrigation water and/or landscape water application; possible floodwaters; canal, reservoir, or pond seepage; surface stormwater; and building roof runoff. Seasonally high water tables can impact the design, efficiency, and operation of the subsurface drain and must be considered.

Surface water should not be introduced into a subsurface drainage system, if at all possible, to prevent the accumulation of debris or other deleterious matter that may cause plugging of the drains and increase maintenance costs. The most cost-efficient system in terms of life-cycle costs may include separate systems to collect and drain surface water and subsurface water.

3.2.7 Soil Information

3.2.7.1 General

Many soil profiles are nonuniform and contain layers of various soil types or rock with varying prop-

erties. Within these strata, the movement of groundwater may vary greatly. Relatively pervious soil and rock formations (aquifers) may exist to influence water movement. An aquiclude (relatively impervious stratum) also affects movement of groundwater.

Permeable soil profiles may not drain freely due to impervious soils or structures near the outer regions; thus, the soil profile at the boundary of areas to be drained must be reviewed. In the soil profile study, this location of an impermeable barrier layer must be established. Subsurface drains must be placed above this layer. The efficiency of the drainage system relates to the distance between the drain invert and impermeable layer. The barrier layer may be nothing more than a change in soil texture within the profile. In practice, the barrier layer is a stratum with permeability less than 10% to 20% of the weighted permeability of all layers above it.

Specific details on soil classification, strata and layers, permeability and drainability, soil-water chemistry, temperature, and area vegetation are important design considerations.

3.2.7.2 Soil Classification

Information on soil classification can be presented in accordance with the Unified Soil Classification System, USDA Textural Classification System, or AASHTO Classification System. The primary concerns are the texture and depth of soil layers, barriers to water movement, apparent permeability and surface infiltration potential, and surface or shallow bedrock. Soil shrink-swell, cracking, subsidence, collapse, consolidation, surface sealing, and compaction potentials are also important factors. The USDA Textural Classification System has some preferred details for use in Section 5, but all systems named previously will help in determining classifications.

3.2.7.3 Strata and Layers

The soil strata and layering require identification because they affect the water movement to and the installation of the subsurface drain lines. Depth, thickness, permeability, slope, and extent of the layers are required data.

3.2.7.4 Permeability and Drainability

These factors should be determined through hydraulic conductivity tests. In situ saturated hydraulic conductivity tests are recommended. Several procedures are presented in the U.S. Bureau of Reclamation, Drainage Manual (1993). Laboratory permeability tests are useful as a screening tool but are not recommended for design of subsurface drains.

3.2.7.5 Soil-Water Chemistry

Salinity, corrosivity, and pH must be determined before selecting pipe and pump materials. Any substances present or projected at the site that could impact either material selection or disposal of the drain effluent must be identified. Substances such as iron pyrites, sodium, calcium, selenium, boron, arsenic, or iron may create disposal and/or material selection problems. Soils with unusually high sodium content are difficult or impossible to drain. Screening tests, including laboratory permeability, pH, settling volume, and electrical conductivity, are useful in detecting the probability of excessive sodium in soils

3.2.7.6 Temperature

An understanding of the temperature range and fluctuations is important for evaluating potential frost action or freeze-thaw. Frost action occurs when moisture is in the freezing zone of the soil and is caused by the freezing and thawing of soil moisture. Frost action may influence material selection and installation depth.

3.2.7.7 Vegetation

Trees and shrubs are particularly important design factors. Water-seeking vegetation, such as the willow tree, is critical in subsurface drain location because the roots tend to plug drains. Other vegetation and landscape details must be noted to provide information on rooting depths, which are important to drain depth and spacing.

3.2.8 Environmental Factors

3.2.8.1 General

Some of the major environmental considerations in the design of subsurface drainage systems follow. These factors must be considered in design to prevent adverse environmental impacts to adjacent parcels of land, residents, and environmentally sensitive ecosystems:

1. Water quality
2. Flooding
3. Wetlands
4. Principal or primary aquifers
5. Hydrology

3.2.8.2 Water Quality

Design, construction, and operation of subsurface drainage systems and components must be performed in such a way as to not contaminate nearby channels. Contamination occurs when subsurface water accumulates excessive minerals, pesticides, or herbicides as the water passes through the soil. Water tables crucial to agricultural farm applications and/or those depended on for sources of potable water must be of high quality. Certain states regulate the possibilities of contamination through required permits for discharge. Such permit requirements must be adhered to in design, construction, and operation. Employing oil-water separation facilities in parking areas and the selective use of discharge points are some of the considerations a designer can use to mitigate possible contamination of water.

3.2.8.3 Flooding

In areas where projects are diverting a high-flow volume of water off-site, it is imperative for the designer to consider the possibility of flooding. Flooding is caused by overflowing streams and runoff from adjacent properties. Potential flooding must also be examined in situations where subsurface drainage is diverted into certain channels such as ditches, enclosed drainage networks, and/or streams. Sufficient capacity for the subsurface flow should be available in the channel during storms. An analysis of downstream channel conditions should be performed to determine any flood hazards. Potential flooding must be adequately considered or liability concerns may develop after the project is constructed.

3.2.8.4 Wetlands

Additional consideration is required for the ecological-environmental aspects of the site when applying artificial drainage to a wetland area.

Wetlands are classified by local, state, and federal governments. Most wetlands are considered to be environmentally sensitive ecosystems. The maintenance of these ecosystems is important. Many wetlands filter natural and man-made pollutants. Changes in the quality and/or quantity of subsurface drainage waters entering a wetland can adversely affect this sensitive filtering process and may cause detrimental effects to flora and fauna associated with the wetland.

Any project associated with a wetland will most likely require permits from local, state, and/or the federal government (Section 9 discusses permits and codes).

3.2.8.5 Principal or Primary Aquifers

These aquifers are often tapped as a main water source. The intended use of this water determines the necessity and amount of protection required. If a possibility exists for contamination of the aquifer, mitigative measures must be taken to prevent such contamination. Other design alternatives, including relocation of the system or a treatment and monitoring program of the subsurface discharge, may be necessary to remove the contamination potential.

3.2.8.6 *Hydrology*

Hydrology describes the movement of water. Development modifies the natural hydrologic cycle and generates an artificial water cycle. Although this has typically been an insignificant design factor, it has grown to be a valid concern in recent years. The development of subsurface drainage systems should follow the natural hydrologic cycle as closely as possible. For example, if the natural cycle exists as rainfall percolating into groundwater, then joining surface watercourses, the man-made cycle should parallel this movement. Not all hydrologic cycles are this simple and thus easily paralleled. It is important to consider the natural or existing hydrologic cycle of the site in the design.

3.2.9 Physical Constraints

Most urban settings have constraints related to existing or planned utilities that must be considered in the design of subsurface drainage systems. Compatibility of proposed systems with existing drain systems is critical to any layout. The location of utilities may require special consideration in the design stage to accommodate pump stations, future development, and master planning.

Additional physical constraints can be identified through a topographic survey, as discussed in Section 3.2.2.

4.0 SYSTEM CONFIGURATION

4.1 GENERAL

An urban subsurface drainage system will include any or all of the following components: collection and conveyance lines, outlets, and appurtenances.

4.2 COLLECTION SYSTEM TYPES

4.2.1 Pipes

Pipes in this standard are intended only for the collection and conveyance of water. Modern pipe materials include concrete, ductile iron, steel, plastic, and clay, and the pipe may have a solid or perforated wall. Pipes are joined in various ways to provide soil-tight or watertight joints. Connector devices include tees, wyes, elbows, and adapters for different diameters and materials.

4.2.2 Geocomposites

Geocomposites of similar or different materials may be used as collection devices to create an "in-plane" envelope for intercepting liquid flowing at right angles to the envelope. Geocomposites may also be used to convey liquid to an outlet. Geocomposites may be installed vertically (i.e., attached to the exterior of a structural foundation wall) or horizontally in a trench.

4.2.3 Geomembranes

Geomembranes may be used as a barrier to liquid flow to waterproof foundation walls, seal under pavements, cutoff fills, and line cuts (e.g., in earth dams, hazardous waste dumps, drainage ditches, and retaining walls). Geomembranes are manufactured in a variety of materials, including plastic, synthetic rubber, and asphaltic compounds.

4.2.4 Geotextiles

Geotextiles may be used as coverings and liners in several construction applications. In drainage applications, geotextiles serve as filters that pass water and colloidal fines while restricting soil migration.

4.2.5 Aggregates

Aggregates may be used as filters or envelopes. Aggregates may be either sands and gravels or crushed stone. Aggregates should be essentially free of sediment and foreign materials.

4.2.6 Wick Drains

Wick drains may be used as vertical drainage to abet the upward flow of water from underground sources and make it free-flowing on the surface. Wick drains accelerate the consolidation of soft and compressible soils.

4.3 CONVEYANCE AND OUTLET

Water collected in a drainage system is normally conveyed to a safe and adequate outlet, such as a natural outfall or storm drainage facility. Where gravity flow is not feasible, pumping is necessary. Most of the collection system types in Section 4.2 may also be used in conveyance systems, which have different design and installation requirements.

4.4 APPLICATIONS

This section includes, but is not limited to, the major applications of the collection system types described in Section 4.2. Combinations of these systems may be used in a drainage system. Geotextiles typically encapsulate aggregate to form a viable drainage system.

4.4.1 Foundation Drains

Foundation drains have application to buildings, bridges, dams, and retaining walls, where structural elements are involved and removal of water is needed. All collection system types have application in foundation drains. In many foundations, a subsurface drainage system using pipes, geocomposites, geomembranes, geotextiles, and aggregates is used. Geomembranes by themselves will keep a structure dry but do not remove the subsurface water. The use of wick drains is limited to areas where the hydraulic head (soil pore pressure) is great enough to force subsurface water to the surface. A wick drain system can have maintenance problems where surface grading is slight and freeze-thaw cycles are a reality. See Figure 4-1 for typical applications of foundation drains.

4.4.2 Roads, Railroads, and Airports

Roads, railroads, and airports have pavement systems with significant length-to-width ratios. The types of drainage systems typically are longitudinal, transverse, and base.

4.4.2.1 Longitudinal Drainage System

A longitudinal system is essentially parallel to the pavement system. It is usually placed at a depth that allows the road section to be gravity drained. Pipe systems, geotextiles, aggregates, and geocomposites are commonly used for longitudinal systems. When curb and gutter are present for a roadway system, discharge

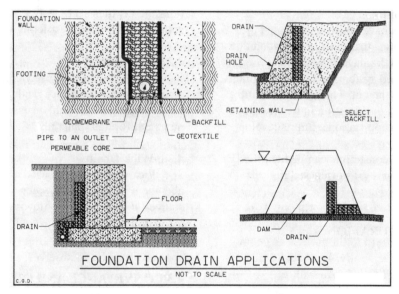

FIGURE 4-1. Foundation Drain Application.

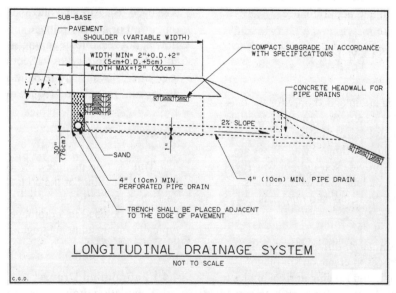

FIGURE 4-2. Longitudinal Drainage System.

into a surface drainage system is usually required. See Figure 4-2 for a typical application of a longitudinal drainage system.

4.4.2.2 Transverse Drainage System

A transverse system is one that drains across the pavement system. It is usually at right angles to the pavement but can be placed at any angle. A transverse drainage system consists of pipe, geotextile, aggregate, and/or geocomposite. The main need for this system is in sections where problems with groundwater are anticipated. This system should connect to the longitudinal subsurface system or the surface drainage sys-

tem. Figures 4-3 and 4-4 show typical transverse drainage systems.

4.4.2.3 Base Drainage System

A base system is one that usually consists of a permeable drainage blanket under the entire pavement system. This system can also add to the structural integrity of the pavement. A properly graded aggregate is often sufficient for a base drainage system. Geotextiles are often used with an aggregate to provide separation and filtration or strength. A properly crowned roadway, railroad, taxiway, and runway subbase will enhance the drainage system. Figure 4-5 shows a typical base drainage system.

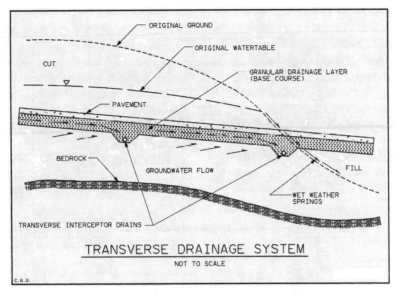

FIGURE 4-3. Base Drainage System.

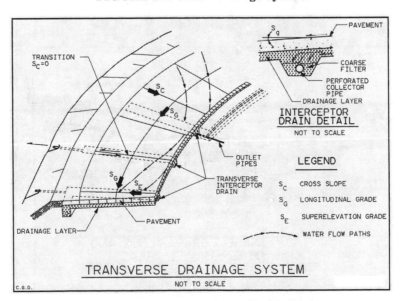

FIGURE 4-4. Transverse Drainage System.

4.4.3 Parking and Other Paved Areas

Parking and other paved areas such as parking lots and play areas have pavement systems where the length to-width ratio approaches 1. The type of drainage needed can vary with the paved surface. Stone and artificial turf will be considered as pervious surfaces. Concrete is considered to be a relatively impervious surface. Asphalt, depending on design, may be pervious or impervious.

4.4.3.1 Pervious Facilities

These facilities can be large with numerous low points where surface and subsurface drainage collect. Accumulated drainage flows through the pervious surface to the subsurface collector system of pipes, geocomposites, and aggregates.

The more pervious the facility, the greater the flow to the subsurface system because infiltration rates are higher. Often these systems are constructed in tandem with a surface drainage system.

Geotextiles and geomembranes can be used depending on effects of the water table. The most important design consideration is to properly design the subbase to get subsurface drainage into the pipe system. Because of the various rates of flow, a thorough study is necessary before the subsurface system is designed for pervious facilities. Figure 4-6 shows a typical pervious system.

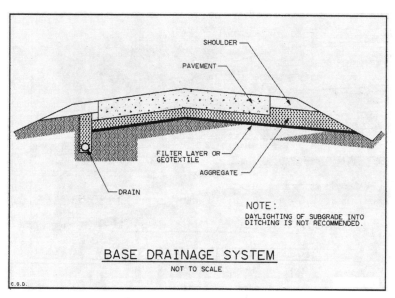

FIGURE 4-5. Base Drainage System.

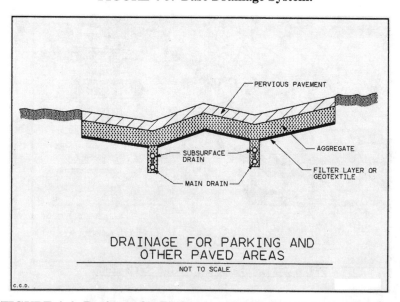

FIGURE 4-6. Drainage for Parking and Other Paved Areas (Pervious).

4.4.3.2 Impervious Facilities

A large impervious facility typically has numerous low points to collect surface and subsurface drainage. However, the volumes of subsurface water are usually very small due to low rates of infiltration through the pavement. Perforated pipes, aggregates, geocomposites, and geotextiles are used in these types of facilities. Unless a rising water table is present, geomembranes, are seldom used because there is rarely a need to keep an area completely dry. Figure 4-7 shows a typical impervious system.

4.4.4 Recreational and Turf Areas

Recreational and turf areas, such as parks, golf courses, and athletic fields, are also used for detention or retention of surface water. Because of the relatively large size of these facilities, the cost of a subsurface drainage system can be an issue. The system used should be designed for low maintenance and should take into account variations of the surface over time. All the collection systems are applicable in one form or another, although the importance of keeping these areas "dry" should be weighed before a system is designed. For example, a revenue-producing golf course must drain well to be back in service as soon as practicable after a storm ends. Keeping adjacent surface waterways free of debris can also be an important part of controlling groundwater levels. Figure 4-8 shows a typical application for a recreational and turf area.

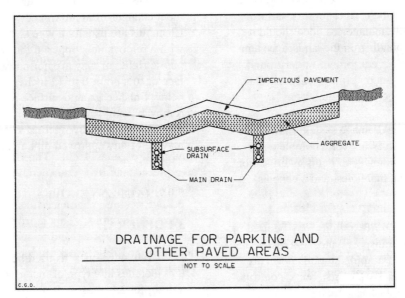

FIGURE 4-7. Drainage for Parking and Other Paved Areas (Impervious).

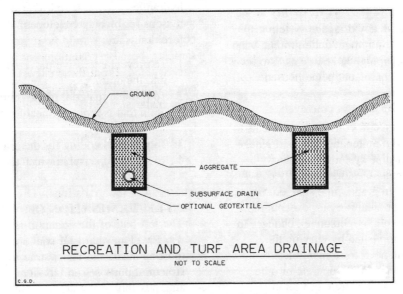

FIGURE 4-8. Recreation and Turf Area Drainage.

4.4.5 Landscaped Areas

Landscape areas are usually situated in urban areas and are typically small areas in relation to the overall development. Large landscape areas are usually part of an overall recreational area. The importance of keeping the areas "dry" must be weighed before a system is designed. For example, an area that has a large amount of foot traffic must be back in service soon after the storm ends.

All collection systems, except wick drains, are applicable. Wick drains are not used because their resultant surface drainage runoff is unacceptable in an urban area. Geomembranes may be used in planter areas to keep foundations dry. Pipe systems may be used in tandem with city drainage systems. Geocomposites, geotextiles, and aggregates may be used to remove surface ponding. The amount of drainage provided should be related to the effect of ponding on the landscaped area.

4.5 APPURTENANCES

An urban subsurface drainage system may include various appurtenances necessary for a complete and operational system, including such items as lift stations, pumping stations, vaults, manholes, and cleanouts.

4.5.1 Pumping Stations

Pumping stations and lift stations may be used in conveyance systems to transport water to a distant and higher discharge outlet. Pumping stations normally include pumps, piping, valves, ducts, vents, controls, electrical equipment, and accessories.

4.5.2 Vaults

Vaults may be used in any drainage system to house electrical or other equipment underground. Vaults normally include ducts, piping, valves, vents, and accessories in addition to the equipment being housed.

4.5.3 Manholes

Manholes may be used in conveyance systems to facilitate inspection and maintenance of the drainage pipe. In small-diameter pipe systems that cannot be entered by personnel, manholes are normally constructed at each change in grade, pipe size, or alignment, and at intervals for cleaning purposes. In pipe systems that can be entered by personnel, changes in alignment may be accomplished by curving the pipeline to eliminate the need for a manhole. Manholes may be constructed of concrete or other approved materials. Precast concrete or prefabricated manhole units are joined in various ways to provide soil-tight joints. Pipe-to-manhole connections should be soil-tight to prevent piping of backfill material into the manhole. The connection should also provide flexibility at the pipe–manhole interface. A watertight connection may be required in some installations. Manholes are normally capped with a metal casting with a removable lid or with a concrete slab that includes the metal casting and removable lid.

4.5.4 Cleanouts

Cleanouts may be used in conveyance systems to facilitate inspection and maintenance of drainage systems that cannot be entered by personnel and are not scheduled for frequent inspection and maintenance. Cleanouts are normally constructed at grade and alignment changes of approximately 45° or greater. Cleanouts are usually a wye section in the pipeline, with a removable stopper in the wye. Cleanouts in public rights-of-way are normally extended to the surface or to a point 6 to 12 inches (150 to 300 mm) below finished ground surface and plugged with a removable stopper. Cleanout wyes should be the same material as the main pipeline. Cleanout extensions may be of any approved pipe material.

5.0 DRAIN ENVELOPES

5.1 GENERAL

Drain envelopes, as used here, is an inclusive term that includes any type of material placed on or around a drain for any reason. Refer to Section 2, Definitions, for more complete definitions of drain envelopes, hydraulic envelopes, and filter envelopes. The primary reasons for placing envelope materials around subsurface drains are:

1. To prevent excessive movement of soil particles into the drain.
2. To provide material in the immediate vicinity of the drain that is more permeable than the surrounding soil.
3. To provide bedding for the drain.
4. To stabilize the soil in which the drain is being placed.

5.2 DETERMINATION OF NEED

Envelope characteristics should be considered in most applications. This is true for all pipe types for structural purposes and to increase water flow rates and, possibly, water storage capacity. Refer to Table 5-1 for guidance.

TABLE 5-1. A Classification to Determine the Need for Drain Filters or Envelopes and Minimum Velocities in Drains

Unified Soil Classification	Soil Description	Filter Recommendation	Envelope Recommendation	Recommendations for Minimum Drain Velocity
SP (fine)	Poorly graded sands, gravelly sands	Filter needed	Not needed where sand and gravel filter is used but may be needed with flexible drain tubing and other type filters	None
SM (fine)	Silty sands, poorly graded sand-silt mixture			
ML	Inorganic silts and very fine sands, rock flour, silty or clayey fine sands with slight plasticity			
MH	Inorganic silts micaceous or diatomaceous fine sandy or silty soils, elastic silts			
GP	Poorly graded gravels, gravel-sand mixtures, little or no fines	Subject to local on-site determination	Not needed where sand and gravel filter is used but may be needed with flexible drain tubing and other type filters	With filter—none Without filter— 1.40 ft/sec (0.43 m/s)
SC	Claycy sands, poorly graded sand-clay mixtures			
GM	Silty gravels, poorly graded gravel-sand silt mixtures			
SM (coarse)	Silty sands, poorly graded sand-silt mixtures			
GC	Clayey gravels, poorly graded gravel-sand-clay mixtures	None	Optional	None—for soils with little or no fines.
CL	Inorganic clays of low to medium plasticity, gravelly clays, sandy clays, silty clays, lean clays			
SP,GP (coarse)	Same as SP and GP.			
GW	Well-graded gravels, gravel-sand mixtures, little or no fines		May be needed with flexible drain tubing	1.40 ft/sec for soils with appreciable fines (0.43 m/s)
SW	Well-graded sands, gravelly sands, little or no fines			
CH	Inorganic, fat clays			
OL	Organic silts and organic silt-clays of low plasticity			
OH	Organic clays of medium to high plasticity			
Pt	Peat			

U.S. Department of Agriculture, Natural Resource Conservation Service, formerly Soil Conservation Service
Soil Conservation Service, 1971

Filters are required in soils with piping or migration potential. Generally, Table 5-1, from the Soil Conservation Service, provides good guidance based on soil classification. Other factors, such as perforation size in the pipe and velocity of flow, must be considered. The filtration requirements take precedence over other considerations for the need for an envelope.

5.3 DESIGN OF ENVELOPES

In selecting a proper filter medium, it is necessary to determine the gradation curve, using wet sieve analysis techniques of the in situ material to be drained. This should be done from actual site samples because gradations for a given soil type can vary widely. From these curves the following design calculations establish limits that are acceptable for filter material. The first criteria proposed by Terzaghi (U.S. Army Corps of Engineers, 1941) for what he termed a filter are:

1. The particle diameter of the D_{15} size of the filter material should be at least 4 times as large as the diameter of the 15% size of the base material (d_{15}). (This would make the filter material roughly more than 10 times as pervious as the base material.)
2. The 15% size of the filter material (D_{15}) should not be more than 4 times as large as the 85% size of the base material (D_{85}). (This would prevent the fine particles of the base material from washing through the filter material.)

The D_{15} size is the particle diameter such that 15% of the material by weight is of a smaller diameter. Also, 85% of the base material, by weight, is smaller than the d_{85} size of the soil. In the case of drainage, the base material is the soil.

The Soil Conservation Service (1991) utilized the filter criteria by Terzaghi as outlined here, along with additional laboratory research (Sherard et al. 1984a, 1984b), and field experience in developing their sand gravel filter envelope criteria (SCS 1991). The NRCS recommends the following gradation limits:

1. Upper limit of D_{100} is 38 mm (1.5 in.).
2. Upper limit of D_{15} is the larger of 7 times D_{85} or 0.6 mm.
3. Lower limit of D_{15} is the larger of 4 times D_{15} or 0.2 mm.
4. Lower limit of D_5 is 0.075 mm (number 200 sieve).

The right half of the equation is for the permeability ratio in the system.

Equation 5-1 displays an acceptable alternative that satisfies both filter and hydraulic envelope requirements when used in conjunction with the Coefficient of Curvature C_c and the Coefficient of Uniformity C_u (USBR 91, ASCE, 1998). The gradations shown are based on soil texture. The C_c must be greater than 1.0 and less than 3.0, and the C_u must be greater than 4.0 with:

$$C_u = \frac{D60}{D10} \qquad (5\text{-}1)$$

and

$$C_c = \frac{(D30)^2}{(D10)(D60)} \qquad (5\text{-}2)$$

where

D10, D30, and D60 = diameter of articles passing the 10%, 30%, and 60% points on the envelope material gradation curve

Not more than 5% of the filter material should pass the #200 sieve or D_5 (filter) > 0.075 mm. This limit is to ensure the permeability of the filter.

The envelope material for the entire drain system should be designed to work with the most restrictive base soil that will be encountered.

5.3.1 Pipe Holes, Slots, and Joints

Envelope materials (or in situ soils if filters are omitted) must be coarse enough not to enter the pipe openings. The following equations should be used:

$$\frac{D_{85}\ (\text{filter})}{\text{maximum opening in pipe}} \geq 1 \qquad (5\text{-}3)$$

5.3.2 Multiple Filters Envelopes

In cases where these ratios cause some difficulty in selecting a filter satisfying both the in situ soil-filter relationship and the filter-pipe relationship, multiple filter layers may be required. In such a case, the recommended limits in Section 6.3. must be met at each interface.

The Federal Highway Administration (1992) provides guidance for the design of the filter or a separator layer to be placed between the base material and the subgrade. A microcomputer program is provided by Wyatt, Baker, and Hall (1998).

5.3.3 Bedding Envelopes

Bedding envelopes may be used to provide proper bedding for sealed or nonperforated drainage pipe. The same design criteria are used for filter envelopes and hydraulic envelopes, but they may be more expensive than necessary if a well-graded coarse sand-gravel material will meet the need.

5.3.4 Hydraulic Envelopes

The design of hydraulic envelopes may be less restrictive than the design of filter envelopes. Sand gradations used for concrete as specified by ASTM C-33 (fine aggregate) or AASHTO M 6-65 will satisfy the NRCS hydraulic envelope criteria and will usually satisfy the NRCS filter envelope requirements for most soils.

5.4 ENVELOPE MATERIALS

5.4.1 Natural Materials

5.4.1.1 Aggregates for drainage envelopes must be selected and sized for maximum permeability and the filter criteria given in Section 5.3. Aggregates must be chemically and structurally stable and must be free of vegetative matter and bentonitic clay films. Flow rates through aggregates vary greatly depending on gradation.

5.4.1.2 Graded sands may be used as filters if their permeability is sufficient and they meet the filter criteria in Section 5.3.

5.4.1.3 Slags

Blast furnace and steel mill slags, although not truly natural, belong in this group. Again, sizing and permeability are critical. These materials must be regarded with caution due to the chemical reaction potential of certain of these slags. In some cases, these materials can be cementitious. They can also release materials, which can clog drains. Structural stability may also be a concern.

5.4.2 Artificial Materials

Geotextiles used in drainage applications should be nonwoven needle-punched, knitted, or spun-bonded. Critical parameters are permittivity, A.O.S. (or E.O.S.), and survivability during construction. Geotextiles are available and are manufactured from polypropylene, polyester, or nylon.

Permittivity values in excess of 1.0 sec^{-1} should be required. Permittivity, not permeability, should be used because it is more important to evaluate the quantity of water that would pass through the fabric under a given head over a given cross-sectional area without regard to fabric thickness.

The A.O.S. of a fabric is the number of the U.S. standard sieve with openings next larger to the actual size of the geotextile openings as determined by the ASTM D4751-95 method. The fabric-soil relationship should be A0S/D85 $\leqq$ 1.

Two properties critical to the survivability of a geotextile are grab strength and elongation. Tested per ASTM D-1682, grab strength should exceed 100 pounds and grab elongation should not exceed 50%.

6.0 HYDRAULICS AND HYDROLOGY

6.1 GENERAL INFORMATION

Successful design of subsurface drainage systems requires an understanding of the behavior of ground-water hydrology and the effects of drainage systems on saturated soils. Furthermore, understanding ground-water collection and discharge properties of drainage systems is essential to ensure desired performance.

6.2 WATER SOURCES

6.2.1 Subsurface Water Sources

In this document, subsurface water is considered to be all water beneath the ground or pavement surface and is sometimes referred to as groundwater.

Soil water is generally of three types: drainable water, plant-available water, and unavailable water. Plant-available water is often referred to as "capillary water" because it is retained by the soil in small soil pores where capillary forces prevent gravity-influenced drainage and is available for plant root absorption.

Drainable water may be considered to be water that readily drains from soil under the influence of gravity. Drainable water moves through soils in direct proportion to the soil's permeability and hydraulic gradient, so low permeabilities result in slow natural drainage of saturated soils.

Unavailable water is held tightly in thin films, surrounding individual soil particles. The strong film bond makes this water nondrainable and unavailable to the vegetation. The amount of this hygroscopic water varies with the surface area of the soil particles and, therefore, is highest in clay and organic soils.

The presence of subsurface water can result from direct surface infiltration or water entering the subsoil from adjacent areas. Another potential contributor to soil wetness is artesian water or water moving up through semipermeable soils under piezometric pressure.

Water infiltration in soils is affected by soil type, season of the year, antecedent moisture conditions, type and extent of vegetative cover, surface "crusting" tendency, and characteristics of the particular rainfall event.

6.2.2 Surface Water Sources

Surface water sources that can contribute to groundwater consist primarily of reservoirs, ponds, canals, open drainage ditches, and rainfall. Often the subsurface drainage requirement is a direct result of seepage from one or more of these sources. The designer of subsurface drainage facilities should investigate any probable surface water sources and, if appropriate, incorporate surface drainage improvements into the design.

In certain situations it may be necessary to introduce surface runoff into subsurface drains, although the practice should be avoided in most cases. When it is necessary, the surface flow inlet must be protected by an effective trash rack, and all downstream pipe should be at least 12 inches (30 mm) in diameter to reduce potential for clogging.

6.3 ESTABLISHING THE NEED FOR SUBSURFACE DRAINAGE

6.3.1 General

Excess soil moisture or ponded surface water may be caused by one or more factors. Disregarding runoff from adjacent areas or subsurface aquifers, low soil permeability is the usual cause of extended water retention following precipitation.

Permeability, or hydraulic conductivity, is used as a measure of the soil's ability to transmit water by gravity. Generally, coarse materials such as sand are highly permeable and have good transmission rates. Clay soils, however, are usually relatively impermeable and water retention is long-term in the absence of a drainage system. Crude estimates can be made of a soil's permeability by realizing that the passage of water depends greatly on voids in the soil structure, explaining chiefly why granular soils with higher void sizes move water better than compact soils with small grain sizes or void sizes.

Soil permeability is best determined by on-site testing. Because of the complex nature of soil composition and its influence on permeability, it is virtually impossible to closely establish this property without direct measurement. While laboratory testing procedures (ASTM D2434) have been established, test accuracy is directly influenced by the inherent difficulties related to duplicating the natural undisturbed soil state. A number of reliable in situ testing methods have been developed for both saturated soils and unsaturated soils as described in the U.S. Bureau of Reclamation Drainage Manual (U.S. Department of the Interior, 1993). At least three in situ tests should be conducted in each major soil group that will influence the drainage system. More tests are recommended for very large systems.

6.3.2 Removal Criteria for Different Environments and Climates

Subsurface drainage, when viewed as water management, is used agriculturally as a method to improve the soil environment for plants and to eliminate operational hazards and nuisances. Benefits in economy and safety are realized.

The rate at which surplus groundwater must be removed relates primarily to the moisture and air requirements of vegetation.

Drainage of urban areas generally relates to lawns, parks, and recreational turf, and usually involves faster removal rates than for agriculture. Faster water removal from the surface and topsoil zone allows quicker access to turf surface and minimizes athletic disturbance of vegetation. Aeration of vegetation is essential to health and durability, and some areas require enhanced drainage for salinity control. Groundwater control is necessary for the correct performance of septic tank absorption fields and for detention ponds.

Climatic conditions must be considered. Soils in humid regions often require more extensive drainage systems than soils in arid regions. Temperature and humidity conditions interact with soil characteristics to influence moisture control requirements.

Comprehensive data for the selection of optimum drawdown rates for nonagricultural drainage systems are not yet available. Local turf specialists should be of assistance, and regional data should identify distinctive climatic conditions such as humidity and evapotranspiration rates.

6.3.3 Special Requirements for Paved Surfaces

Vehicular traffic on pavements with saturated subbase results in early deterioration. Water enters through openings in the pavement from high groundwater conditions or as seepage from adjacent regions. Correctly designed paved areas should have highly permeable base or subbase construction and drainage systems for promoting rapid outflow of infiltrated water.

6.4 BASIC SUBSURFACE DRAINAGE THEORY

6.4.1 Controlled Water Table Elevation

Subsurface drainage is accomplished by placing an artificial channel below the water table so that the hydraulic head of the channel is less than that of the soil to be drained. The hydraulic head differential creates a hydraulic gradient in the direction of the artificial channel, depressing the phreatic line (also called free water surface; see Figure 6-1) in the vicinity of the artificial channel. The constant removal of water flowing into the drainage sink maintains the hydraulic head differential, thus maintaining the depressed phreatic line.

The hydraulic gradient and the hydraulic conductivity of the soils to be drained govern the rate at which water moves toward the sink. Control of water is accomplished by controlling the hydraulic gradient. Therefore, flow is regulated by adjusting the depth of the sink and the spacing between sinks, and by locating the sinks to take advantage of the more permeable soils in the area to be drained.

6.4.2 Drainable Versus Nondrainable Groundwater

Drainable water, or "gravity water," is that which is free to move through the soil by the force of gravity or hydraulic gradients. When all voids are filled with water, the soil is considered to be saturated. Some water is held in the soil against gravity. It includes the film of water left around the soil grains and the water filling the smaller pores after gravity water has drained off. This water is considered to be nondrainable groundwater.

6.4.3 Drainage Formulas and Recommended Practices

Selection of the proper drainage system to control the anticipated flow will vary with the specific application, but in all applications the procedure begins with a determination of the allowable subsurface water elevation and the drain depth. Spacing of drains can then be calculated and a discharge can be computed for each drain.

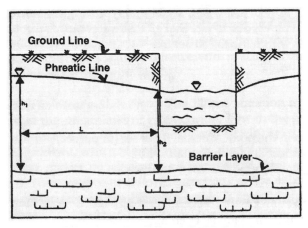

FIGURE 6-1. Soil Water Movement.

Most methods for estimating drain spacing are empirical and were developed to meet specific characteristics of a particular area. Some methods are based on assumptions of steady-state flow conditions where the hydraulic head does not vary with time. Other methods assume transient flow conditions where the hydraulic head changes with time.

6.4.3.1 Steady State

The most widely used steady-state formula for subsurface drainage spacing is the ellipse equation. In using the ellipse equation, the depth to the drain is first established, then spacing of the drains is computed by the formula. The ellipse equation is based on an assumption that the streamlines of flow in a gravity drainage system are horizontal and the velocity of flow is proportional to the hydraulic gradient or the free water surface. Although approximate, these assumptions may closely approach actual conditions in certain sites. For this reason, use of the ellipse equation should be limited to the following conditions:

1. Where recharge is primarily from the soil surface.
2. Where the groundwater flow is known to be largely in a horizontal direction, such as stratified soils with relatively permeable layers acting as horizontal aquifers.
3. Where soil and subsoil materials are underlain by a barrier layer at relatively shallow depths (less than twice the depth to the drain), which restricts vertical flow and forces the groundwater to flow horizontally toward the drain.
4. Where open ditches are used or where drains with sand and gravel filters or porous trench backfill are used. These conditions cause a minimum of restriction to flow into the drains and minimize convergence of flow at the drains.

The ellipse equation (ASCE 1998) is usually expressed in the form:

$$S = \sqrt{\frac{4K_s(m^2 - a^2)}{q}} \qquad (6\text{-}1)$$

where

S = drain spacing, ft (m)
K_s = average hydraulic conductivity, in/hr (m/day)
m = vertical distance, after drawdown, of water table above the barrier at midpoint between drain lines, ft (m)
a = depth of barrier layer below drain, ft (m);
q = drainage coefficient (steady-state surface recharge rate), in./hr (m/day)

Details of the symbols and parameters for Eq. 6-1 are shown in Figure 6-2. A correction for convergence losses should be used. A convergence correction developed by Hooghoudt USBR 1993) considers the loss in head required to overcome convergence in the primary spacing calculation. The calculations derive an equivalent depth to barrier, which is then inserted into the ellipse equation in place of the actual depth to barrier, and a new spacing is calculated. The procedure for computing the adjustment for convergence is discussed in U.S. Bureau of Reclamation and U.S. Department of Agriculture references (see Section 3) with examples presented for use of both steady-state computations and transient-state computations.

6.4.3.2 Transient State

Drainage formulas have been developed, the most notable of which are those developed by Ernst, Hooghoudt, Dumm, and Van Schilfgaarde. These drainage models are considerably more complicated and apply to more complex types of geologic and hydrologic conditions. The transient-state drain spacing processes are more detailed in that they require the designer to develop a pattern of recharge events to simulate anticipated future conditions. In return, the processes facilitate analyses of the drain design for economic assessment and risk assessment. Economic assessment allows the designer to select the most economical depth and spacing, while the risk assessment reveals the probability that the water table will rise above design limits with any given design. Transient-state spacing computations should be corrected for convergence using the same methods described for steady-state computations.

The particular formula selected for computing the spacing of relief drains is influenced by site conditions and experience from drains in similar soils and climates.

Reference is made to background material presented by TeKrony, Sanders, and Cummins (March/April 2004).

6.5 SUBSURFACE DRAIN APPLICATIONS

6.5.1 Soil Consolidation Using Wick Drains

Compressible soils with high water content are generally unsuitable on construction sites selected for any structure requiring a stable foundation. Since the damaging effects of soil settlement are well documented, the practice of soil stabilization through the process of consolidation has been widely accepted. This process fundamentally involves the development of a surcharge loading on the site to produce excess pore pressures in the water, forcing water from the soil pores. According to the theory of consolidation, the results vary inversely with the soil permeability and directly to the square of the longest drainage path. In most cases, especially where fine silts and clays are present, the process is extremely long if unaided by a drainage system.

Establishing vertical trenches and filling them with sand or inserting geocomposite wicks into the soil to be consolidated shortens the drainage path and accelerates the rate of consolidation. The design of sand drain systems is commonly being replaced by smaller geocomposite wicks. This permits quicker and more reliable construction.

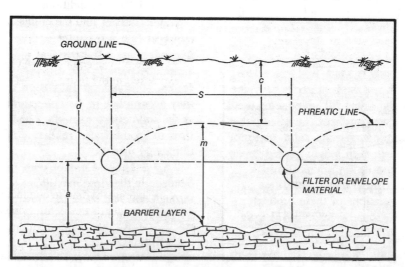

FIGURE 6-2. Cross-Sectional View Showing Symbols Used in Ellipse Equation.

6.5.2 Groundwater Removal with Relief Drains

Relief drains are broadly considered to be any product or construction that accelerates the removal of drainable subsurface water. Such drains are commonly required in soils whose properties result in extended saturation periods following rainfall events, thus affecting infiltration and runoff, vegetative health, and seepage forces on underground structures. In the broadest sense, relief drains can take the form of trenches or paved channels, but they are more commonly a buried product or structure that is systematically designed and located to serve as a subsurface water collector or serve as both a surface water collector and a subsurface water collector. Relief drains also discharge the collected water to a selected area or structure.

Relief drain design and construction has evolved over many centuries to include sand- and aggregate-filled trenches (both with and without perforated pipe in the lower portion of the select fill), pipe with small perforations through the wall surfaces (installed either with or without a geotextile covering), solid-wall pipe installed with gaps between each length (called "open joint pipe"), and geocomposite fin drains (formed plastic interior members overwrapped with filter textiles). Figure 6-3 is one configuration of these products and systems that provide groundwater removal from turf areas and protect subsurface structures from seepage forces.

Drainage systems must serve the two primary functions over long periods of time: Collect groundwater and then discharge the collected water to a selected point. In nearly all cases, gravitational force and the resulting driving energies are relatively small, requiring care in design and placement to avoid major performance reduction.

6.5.3 Seepage Control with Interceptor Drains

Seepage is the slow movement of gravitational water through soil and rock; the flowing water creates a frictional drag force proportional to flow velocity. Seepage forces have been significant contributors to the failures of dams and other earthen structures, in addition to subsurface structures such as foundations and retaining walls.

In broad terms, there are three ways to deal with seepage groundwater. In a few cases, the structure can be designed to allow passage of groundwater without endangering the integrity and performance of the structure. The remaining two seepage control methods either keep the water out or use drainage methods to control water removal. The techniques for shielding a structure involve wall or curtain construction.

Controlling seepage via drainage techniques essentially involves the placement of a subsurface drain so that it intercepts the groundwater and protects the structure, which otherwise would be affected by the seepage force. As shown in Figure 6-3, the general design of such interceptor drains is normally perpendicular to the slope of the water table. A good example of the use and effectiveness of interceptor drains is the long-standing practice of surrounding building foundations with highly permeable materials such as sand or aggregate. Retaining wall designs also frequently have nearly full-height drainage material to intercept seepage water. A suitable outlet must be provided for the collected water.

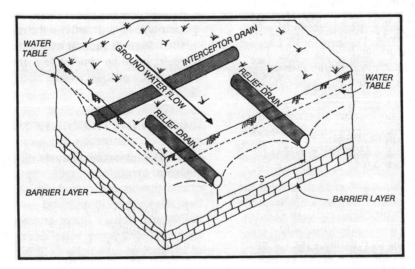

FIGURE 6-3. Isometric Profiles of Relief and Interceptor Drains.

There are many design similarities between relief drains and interceptor drains. Likewise, the precautions regarding long-term performance are similar. In both cases, extreme care is required to balance discharge capacity with inflow expectations and, furthermore, to evaluate the soil environment and inflow area materials to protect against reductions of inflow or discharge capacities.

Also, similar to subsurface relief drains, geocomposites of sufficient size and geometric shape may be employed as interceptors, and these are particularly adaptable to rigid flat surfaces such as retaining walls and foundations.

6.5.4 Water Control of Pavement Structural Sections

Concrete and asphalt areas (including highways, streets, parking lots, and parks) usually are supported by an underlay of sand or stone, which should provide both structural stability and drainage for the total structure. However, significant amounts of water can enter this base or subbase course from a variety of sources and, without the opportunity to drain, remain entrapped for extended time periods. Combinations of vehicular loadings and freeze-thaw cycles on pavements with saturated base layers cause large hydraulic and ice pressure forces that accelerate breakdown of the pavement. Such deterioration is widely experienced on heavily traveled roadways and is considered the major cause of deterioration on most highway systems in the United States. Other paved areas also suffer shortened lives due to inadequate subgrade drainage. Included in this group are airport runways and taxiways, residential streets, parking lots, sidewalks, and tennis courts. Any outdoor pavement should be designed for rapid removal of infiltrated water in the base course following rainfall. Federal Highway Administration; (FHWA 1992) training materials suggest that the saturation level be reduced to 85% or less within 2 hours after rainfall ceases.

The design of pavement drainage generally involves both the base layer and a specific drainage product or system. Designs require a drainable structural base or subbase course, employing open-graded materials. Low-permeability sand bases, except for small areas with light loadings (nonvehicular), may be inadequate.

Open-graded permeable structural courses slope toward the drainage system, where water from the course is collected and discharged. Conventional roadway design is to slope the course following the roadway center crown or elevation and place a collector drain system at the downslope edge, which may be

both sides of a relatively flat roadway or one side of a superelevated curve. Once the aggregate layer drains freely into the edge drain(s), the drain must function as a relief drain of significantly greater length. Because of the long runs in typical edge drain installation, side outletting into a drainage ditch or storm drain is necessary for outflow-inflow continuity. Outlet spacing is typically several hundred feet (multiples of 30 meters) on centers for pipe or high-column edge drain systems on major highways. Catch basins provide a convenient opportunity to direct flow from the subsurface edge drain into the storm drain system.

6.6 INFLOW-OUTFLOW CONTINUITY

6.6.1 Flow Continuity for Maximum Performance

For adequate subsurface drainage system performance in any application, designs balancing outflow and inflow are essential. All sources of inflow must be identified and quantified. Inflow sources are most commonly: (1) surface infiltration (rainfall, snowfall, and irrigation sources); (2) water transfer from adjacent areas (springs, waterfalls, and watersheds); and (3) high water table.

The flow of water through soil and/or pervious base material into subsurface collectors, accompanied by discharge and disposal, is dependent on several factors: (1) soil permeability (hydraulic conductivity); (2) flow energy losses at collectors (such as siltation of filter envelopes or fabric, convergence of flow lines, entrance resistance of perforations); and (3) hydraulic characteristics of the subsurface drain system.

Soil permeability is carefully evaluated because it is a requisite element to subsurface drainage performance. Flow reduction potential due to convergence and siltation must be considered because these also contribute significantly to the result. Equally important is the determination of hydraulic properties for the collector and drain systems to permit unrestricted discharge of collected water.

6.7 INFLOW TO COLLECTORS

The following methods should be used for drain sizing.

6.7.1 Relief Drains

Procedures for calculating the flows to relief drains involve prior computed drain spacings. The spacing is computed using Eq. 6-1 as described in Section 6.4. The area served by parallel relief drains is equal to the spacing times the length of the drain plus one-half the spacing. The total discharge can be expressed by the steady-state equation:

$$Q_r = \frac{qS(L + S/2)}{43,200} \qquad (6\text{-}2)$$

(USDA 1969 and 1971) (ASCE 1998)

where

Q_r = relief drain discharge, cubic feet per second
q = drainage coefficient, inches per hour
S = drain spacing, feet
L = drain length, feet or SI units

$$Q_r = qS(L + S/2) \qquad (6\text{-}3)$$

where

Q_r = relief drain discharge, cubic meters per day
q = drainage coefficient, meters per day
S = drain spacing, meters
L = drain length, meters

Figure 6-4 shows a parallel relief drain system; the shaded area indicates the area served by one of the relief drains. The value of Q_r is for the drain in its entirety (i.e., at its discharge point to a collector drain). The flow into the relief drain will be uniform along its length, with the discharge at any point along the drain proportional to the length. This suggests that collector drains receiving flow from several relief drains can be sized according to the flow carried at the various points along the collector. Depending on the size, shape, and configuration of the area to be drained, this variable sizing could result in significant cost savings.

The inflow to drains may also be calculated using transient-state methods. The transient-state equations require a known depth to groundwater, a known depth to barrier, and a known hydraulic conductivity, all of which can be directly measured.

It does not require the drainage coefficient, which cannot be measured directly. The transient-state equations may be used to verify the estimated drainage coefficient by comparing the results from the two methods, steady-state and transient-state, thereby lending credence to the drain spacing.

When a single drain line is used rather than a system of spaced drains, regional groundwater flow may account for a large part of the flow to the drain, and the contributing area is usually not known. In these cases, Eqs. 6-4 or 6-5, which do not require a known contributing area, can be used to determine the inflow rate to the drain.

The transient state equations are: (USBR 1993)

$$q_p = \frac{2\pi K y_0 D}{86,400\ L} \qquad \text{(Eq. 6-4, for drains above the barrier)}$$

and

$$q_p = \frac{4KH^2}{86,400\ L} \qquad \text{(Eq. 6-5, for drains on the barrier)}$$

where

q_p = discharge from two sides per unit length, in cubic feet (cubic meters) per second per foot (meter)
y_0 or H = maximum height of water table above the drain invert in feet (meters)
K = weighted average hydraulic conductivity of the soil profile between the water table and the barrier in feet (meters) per day
D = average flow depth ($D = d + y_0/2$) in feet (meters)
d = the distance from drain to barrier in feet (meters)
L = drain spacing in feet (meters)

Reference is made to background material presented by TeKrony, Sanders, and Cummins (March/April 2004).

6.7.2 Wick Drains and Chimney Drains

Groundwater moves into the drain from the surrounding soils due to higher pore pressures; soil consolidation rates are accelerated by the escape of water from the soil voids. Volume flow rate to these drains is a function of the number and type of alternating soils in the profile; the permeability and porosity of the soil; extent of soil disturbance during

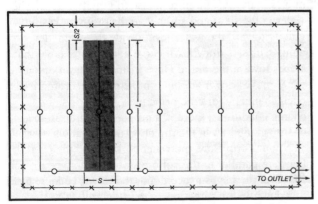

FIGURE 6-4. Plan View of Relief Drain System.

installation of the drains; the radial flow resistance of the drain; blinding/clogging potential of the sand or geotextile wrap; and, in the case of geocomposite drains, the amount of free (open) inflow surface; the effective size of the drain; and the number of drains installed.

6.7.3 Interceptor Drains

In calculating the flow to an interceptor drain from upslope sources, the Darcy equation may be used. The flow per unit length into an interceptor drain may be calculated using the equation: (USBR 1993)

$$q_u = K_c IA(y/(y + d)) \qquad (6\text{-}6)$$

where

q_u = volume rate of flow into drain per unit length, cubic feet per second per foot (cubic meters per second per meter)

K_c = weighted average unit hydraulic conductivity of the strata above the barrier, feet per second (meters per second)

I = slope normal to groundwater contours, feet per foot (meter per meter)

A = saturated area in square feet per foot (square meters per meter) of flow in a unit length of drain

y = height of maximum water surface immediately above proposed drain, feet (meters)

d = distance from drain invert to impermeable barrier, feet (meters).

K_c is computed from:

$$K_c = \frac{K_1 T_1 + K_2 T_2 + \ldots + K_n T_n}{T_1 + T_2 + \ldots + T_n} \qquad (6\text{-}7)$$

where

K_i = hydraulic unit conductivity for layers K_1 through K_n of the soil profile, feet per second (meters per second)

T_i = thickness of layers T_1 through T_n, feet (meters).

Generally, the maximum water table height would be used to obtain the saturated depth from which K_c is obtained. This same depth would be used to obtain the area A for a unit width. The plane along which the area must be obtained is parallel to the contours or perpendicular to the direction of flow.

6.7.4 Perimeter Drains and Edge Drains

Perimeter drains and edge drains function similarly to relief drains. Used longitudinally on roads and airport runways, edge drains are designed to quickly remove excess water from the base or subbase course and subgrades of pavement structures to prevent premature deterioration of the pavement.

Flow into an edge drain or a perimeter drain is similar to that of the interceptor drains discussed previously, except most flow enters from the pavement side of the drain. Side outlets for edge drains must be spaced according to the estimated flow in the edge drain and the hydraulic flow capacity of the drain. In using some of the newer geocomposite fin drains, laboratory tests for inflow and capacity of each specific product being evaluated must be used for comparisons of appropriateness since specific criteria have not been adopted by consensus.

6.8 HYDRAULICS OF SUBSURFACE DRAINS

6.8.1 Outflow of Collected Water

The design of a subsurface drainage system includes the hydraulic design of all conduits that collect the excess water from the soil and deliver the water to the outlet for safe disposal. The hydraulic capacity of the drains should be checked at all points of size modification and at lateral connections. Subsurface drains are normally designed as open channels with a free water surface (at atmospheric pressure) inside the pipes. The amount of water collected depends on the drainage coefficient for the area and the size of the area contributing water to the drain. If lateral seepage is occurring from outside the area drained, the conduits must be large enough to carry the additional water that will enter the system.

6.8.2 Fundamental Hydraulic Theory for Drains

The specification of a free water surface inside the drainpipe as a design factor means that all water moves in the system in response to gravity forces. The water in the saturated soil above the drain moves by gravity toward the water level inside the drain, developing a classic drawdown curve and a water table that slopes toward the drain. Once inside the drain, the water moves down gradient in the pipe according to open channel hydraulic principles. Under free discharge conditions, there is a progressive drop in the energy grade line in the direction of the outlets. For steady uniform flow, the water surface and the energy lines are parallel to the bottom slope of the pipe. However, in a subsurface drain, the flow rate increases in a downstream direction so the flow rate to be carried by the pipe at any point along its length is not a constant and the slope of the energy line also varies along the length of the pipe. Drainpipes are usually installed on a uniform grade. The maximum carrying capacity for the drain is calculated at the end of each section having a uniform pipe size and a constant bottom slope. Consider energy losses due to friction, changes in flow direction, changes in the size or shape of the pipe, control or measurement devices, inlet and outlet head losses, and momentum transfer. Momentum transfer required to accelerate water entering the pipe at nonzero angles to the flow direction in the pipe must also be considered at junctions.

An adequate pipe bottom slope and adequate minimum flow velocity are important in subsurface drains to minimize sediment deposition. Slopes of at least 0.001 are usually adequate. Excessively steep slopes may cause erosion of the soil outside the pipe perforations and result in failure of the drain. Some drains, such as foundation drains, may have to be laid with a zero slope. However, a hydraulic grade line will develop to remove the water as long as an adequate outlet is available and the pipe is properly designed.

The Manning equation discussed in Section 6.8.3 is one method used to estimate pipe carrying capacity.

A spatially varied flow calculation using numerical methods may be made using the ordinary differential equations for spatially varied flow (Chow 1959) to ensure that the hydraulic grade line will remain inside the conduit and that the design is economical. In the latter case, the Manning equation is used to calculate a friction slope at each computational section along the conduit. Figure 6-5, based on Graber (2004), provides a chart that can be used for design of circular conduits of single pipe diameter with uniform inflow and free discharge conditions (tailwater elevation less than downstream critical depth elevation). The variables in Figure 6-5 are as follows:

n = Manning roughness coefficient
d_o = pipe diameter (m, ft)
L = pipe length (m, ft)
Q_L = rate of flow (m^3/sec, cfs)
g = gravitational acceleration (9.81 m/sec^2, 32.2 ft/sec^2)
K_n = 1 for SI units, 1.486 for customary units

For more complex systems, involving, for example, changes in pipe diameter or downstream submergence, a computer program may be used for design and analysis purposes (Graber 2004).

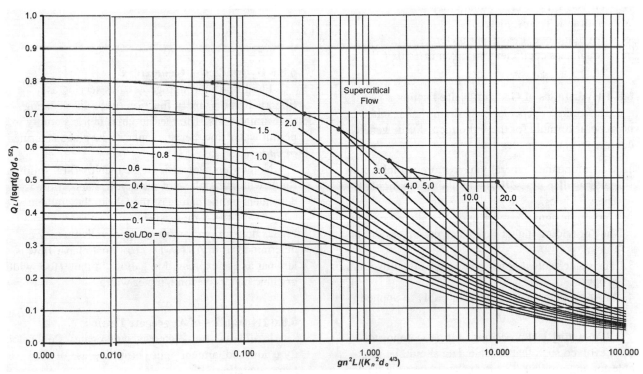

FIGURE 6-5. Design of Circular Conduits of Single Pipe Diameter with Uniform Inflow and Free Discharge Conditions.

6.8.3 Manning Equation

The following equation was proposed by Manning in 1889 for open-channel flows:

$$Q = \frac{A\,1.49\,R^{2/3}S^{1/2}}{n} \qquad (6\text{-}8)$$

where

Q = flow rate, cubic feet per second

A = cross-sectional area of the water stream, square feet;

R = hydraulic radius: $R = A/WP$, feet;

S = friction slope or slope of the energy grade line, feet per foot (which, as discussed above, equals the slope of the drain only for steady uniform flow)

WP = wetted perimeter, which is the length of the periphery of the cross-sectional shape of the liquid in contact with the pipe wall surface, feet

n = Manning roughness coefficient for the drain product (refer to Table 6-1) or

$$Q = \frac{A\,R^{2/3}\,S^{1/2}}{n} \qquad (6\text{-}9)$$

where

Q = flow rate, cubic meters per second

A = cross-sectional area, square meters

R = hydraulic radius, meters

S = friction slope, meters per meter

n = roughness coefficient, as given in Table 6-1

6.8.4 Hydraulics of Geocomposite Drains

Research on geocomposite drains has resulted in no general formula for these products. Some general comments can be made:

1. Geocomposites generally have more complex flow patterns than pipes, thus contributing to higher flow energy losses.
2. Certain designs are more hydraulically efficient, thus providing higher flow rates per unit width or thickness.
3. The outer geotextile wrapping may affect hydraulic properties.
4. Most geocomposites react noticeably to applied earth loads.

Laboratory test procedures for geocomposites have evolved such that reliable data should be available for any product. Test data should provide flow rates in a product test situation modeling the in-service environment.

Table 6-1. Recommended Design Values of Manning Roughness Coefficients for Closed Conduits and Open Channels (ASCE, 1982 and ASCE, 1993)

Conduit Material	Manning's n
Closed conduits	
Asbestos-cement pipe	0.011–0.015
Brick	0.013–0.017
Cast iron pipe	
Cement-lined and seal-coated	0.011–0.015
Concrete (monolithic)	
Smooth forms	0.012–0.014
Rough forms	0.015–0.017
Concrete pipe	0.011–0.015
Corrugated-metal pipe	
Plain annular	0.022–0.027
Plain helical	0.011–0.023
Paved invert	0.018–0.022
Spun asphalt lined	0.011–0.015
Spiral rib metal pipe (smooth)	0.012–0.015
Ductile iron pipe (cement lined)	0.011–0.014
Plastic pipe (corrugated)	
3–8 in. (75–200 mm) diameter	0.014–0.016
10–12 in. (250–300 mm) diameter	0.016–0.018
Larger than 12 in. (300 mm) diameter	0.019–0.021
Plastic pipe (smooth interior)	0.010–0.013
Vitrified clay	
Pipes	0.011–0.015
Liner plates	0.013–0.017

6.8.5 Hydraulics of Geotextiles

Fluid flows through geotextiles serving as a filter, with the primary function of the fabric being to restrain solids. The measure of a fabric's ability to permit water flow perpendicular to its plane is termed *permittivity*.

Nonwoven geotextiles are capable of transmitting fluids within the plane of the fabric. Planar flow is dependent on the nature of the fabric, the gradient (slope), and the compressive forces on the material. Planar flow properties for any geosynthetic may be determined from ASTM D4716. Most planar flow is laminar and is expressed in gallons/minute/foot width or cubic meters/second/meter width.

6.8.6 Hydraulics of Aggregate Drains

In all but extremely coarse aggregates, flow velocity is low and laminar, thus permitting use of the basic Darcy equation:

$$Q = K\,i\,A \qquad (6\text{-}10)$$

where

Q = flow rate, cubic feet per second (cubic meters per second)

K = permeability of aggregate, feet per second (meters per second)

i = hydraulic gradient (approximately equal to the slope ($\Delta h/l$), feet per foot (meter per meter)

A = cross-sectional area of the aggregate-filled trench, square feet (square meters)

The most significant variable is permeability, which is principally dependent on the voids established by the aggregate's particle size and gradation. Aggregate permeability can vary widely. Laboratory testing to confirm permeability of selected aggregate versus the design requirements is highly recommended.

The amount of fines in the aggregate mix affects permeability. Generally, higher fine content results in lower permeability, indicating the desirability of clean, washed stone. Safeguards are required to minimize passage of excessive fines from adjacent soil into the aggregate drain.

6.8.7 Hydraulics of Consolidation Drains

Flow within consolidation drains is proportional to the cross-sectional area and the hydraulic conductivity of the drain fill material in the case of sand drains, and flow energy is dissipated in moving the water column upward. The vertical drain must not develop significant restrictions to flow within the drain. Head losses in vertical drains are determined by:

$$h_d = \frac{q_d d}{2 k A} \qquad (6\text{-}11)$$

where

h_d = total head loss in drain, feet (meters)

q_d = groundwater discharge for each drain, cubic feet per second (cubic meters per second)

d = depth of the drain, feet (meters)

k = permeability coefficient of drain backfill, feet per second (meters per second)

A = cross-sectional area of the drain per running foot, square feet per foot (square meters per meter)

Note that higher drain backfill permeabilities and larger cross-sectional areas reduce head losses proportionally. As consolidation progresses, the density of the drain backfill material may increase and there may be an accompanying decrease in permeability.

Geocomposite vertical drains may be evaluated in a fashion similar to sand drains. In determining head loss, the hydraulic transmissivity (flow rate potential) of the wick drain (determined by laboratory test) is substituted for the permeability coefficient and the cross-sectional area in Equation 6-11.

Actual pipeline performance depends on the effects of abrasion, corrosion, deflection, alignment, joint condition, and sedimentation.

7.0 STRUCTURAL CONSIDERATIONS

7.1 LOADING

7.1.1 General

The loads that may be applied to an urban subsurface drainage system are categorized as either dead loads or live loads. Equipment live loads may be critical during construction operations.

Pipe and geocomposite systems are designed to carry live loads and dead loads. Geomembrane, geotextile, and aggregate systems, by their nature, normally do not have to be designed for structural strength to carry live loads or dead loads, but must be designed for stability and possible shear forces during construction.

7.1.2 Dead Loads

Dead loads may include earth loads, weight of pipe, weight of fluid in the system, building foundation loads, and surcharge loads. The primary dead load, which usually governs, is the earth load. The magnitude of the earth load is dependent on the unit weight of the soil and depth of the fill and can be determined by the design methods found in the following references: American Association of State of Highway and Transportation Officials (AASTO), 1992; American Concrete Pipe Association, 1988 and 1992; American Iron and Steel Institute., 1971; National Clay Pipe Institute, 1982; Spangler, 1966; and U.S. Department of Agriculture, 1958 and 1980.

7.1.3 Live Loads

Live loads are imposed by highway vehicles, trains, aircraft, and construction equipment. Design live loads are typically taken as the AASHTO H-20, HS-20, or alternate interstate loading for highways and the Cooper E 80 loading for railroads. For distribution of these loadings onto the buried structure, see AASHTO, 1992, for highway loadings; American Railway Engineering Association, 1993, for railroad loadings; and U.S. Department of Transportation, 1970, for aircraft loadings.

7.1.4 Construction Loads

Construction loads from heavy construction equipment traveling over or across an installed drain system may create load concentrations in excess of design loads that may displace and damage the system. Such locations should be evaluated by the engineer to determine if displacement and damage may occur. If necessary, crossing location requirements should be detailed in the construction contract documents. Although crossings for pipe systems should be evaluated on a case-by-case basis, a common crossing consists of a temporary earth fill constructed to an elevation of at least 3 feet (1 m) over the top of the system and to a width sufficient to prevent lateral displacement of the system.

7.2 EMBEDMENT

The structural performance of buried pipe is dependent on the interaction between the soil and the pipe. Therefore, the pipe embedment must be selected for structural and drainage characteristics. Structural characteristics of the embedment include consideration of the dimensions of the embedment around the pipe; the type of soil, density, potential for live loads, and compaction of the embedment, native soil, and fill; the depth of burial of the pipe; and the height and characteristics of the water table. The required dimensions and type of soil, density, and compaction of the embedment are dependent on the pipe stiffness.

Flexible pipes, such as plastic and corrugated metal, use the embedment materials to transfer vertical loads into the adjacent soil. Rigid pipes, such as concrete and clay, transfer vertical loads directly into the bedding with minimal load transfer into the adjacent soil. Therefore, the required structural characteristics of the embedment vary with the type of pipe and shall be determined in accordance with Section 7.3.

The type of equipment to be used to compact the embedment and fill should be evaluated by the engineer to determine if pipe damage and displacement may result from its use. If necessary, equipment limitations should be detailed in the construction contract documents.

7.3 PIPE DESIGN

There are several pipe design methods in existence. The appropriate method depends on the pipe application or end-use, the type of pipe material, and possibly the project authority or owner. Acceptable design methods for pipe projects within the scope of this practice are presented for each pipe material in Sections 7.3.1 through 7.3.4.

The trench bottom should have a 120° semicircular groove cut in it to cradle the pipe and provide

structural support. The groove should be cut, not pressed, in the trench bottom. The groove can also be a 90° V-groove. The groove can be cut into the bottom of the trench by the pipe feeding boot or by the cutting blades on the digging apparatus. If the bottom of the trench is filled with fine gravel, which surrounds the pipe, the groove is not necessary.

7.3.1 Corrugated Metal Pipe

Service load and load factor design methods and embedment requirements are presented in Section 12 of *Standard Specifications for Highway Bridges* (American Association of State Highway and Transportation Officials) and in *Handbook of Steel Drainage and Highway Construction Products* (American Iron and Steel Institute).

7.3.2 Thermoplastic Pipe

Manufacturers' literature should be checked for recommended design methods and embedment requirements. Other sources of design information are listed in the following paragraphs.

Handbook of PVC Pipe: Design and Construction (Uni-Bell PVC Pipe Association, 1986) is a unified source for PVC pipe design and construction.

ASCE Manual No. 60, "Gravity Sanitary Sewer Design and Construction" (American Society of Civil Engineers, 1982), presents design methods and requirements for all thermoplastic pipes.

NCHR Report 225, "Plastic Pipe for Subsurface Drainage of Transportation Facilities" (Chambers et al., 1980), presents a state-of-the-art review of thermoplastic pipe design methods.

Structural Design Methods for Corrugated Polyethylene Pipe published by the Corrugated Polyethylene Pipe Association presents information on design of corrugated polyethylene pipe.

Section 18 of the AASHTO *Standard Specifications for Highway Bridges* presents service load and load factor design methods and embedment requirements for some PVC and PE pipe products.

ASTM D-2321-00 Standard Practice for Underground Installation of Thermoplastic Pipe for Sewers and Other Gravity-Flow Applications.

7.3.3 Precast Concrete Pipe

Service load and load factor design methods and embedment requirements are presented in Section 17 of *Standard Specifications for Highway Bridges* (AASHTO). The American Concrete Pipe Association's *Concrete Pipe Design Manual* presents data and information on the design of concrete pipe systems. The design manual is a companion volume to the *Concrete Pipe Handbook*, which provides an up-to-date compilation of the concepts and theories that form the basis for the design and installation of precast concrete pipe. American Society of Civil Engineers Standard 15 (ASCE 1998) also contains valuable information on this subject.

7.3.4 Clay Pipe

Service load design methods and embedment requirements are presented in *Clay Pipe Engineering Manual* (National Clay Pipe Institute).

7.4 OTHER SYSTEMS

Geomembranes, geotextiles, and aggregate systems, by their nature, do not have to be designed for structural strength to carry live loads or dead loads. These systems, however, must be designed for stability, for possible shear forces from construction operations, and to maintain maximum flow capacities after placement. Geomembranes and geotextiles should have sufficient strength to prevent tearing during construction and operation.

Geocomposites should be of such shape and have sufficient strength to prevent major deformations of the geotextile into the core that will reduce flow capacity under load. For pavement applications, geocomposite drains should have sufficient strength to withstand a long-term compressive force of 20 psi (138 kPa) normal to the plane of the drain or twice the active soil pressure at the drain depth, which-ever is greater. For other applications, geocomposite drains must be evaluated against the anticipated sustained dead loads to ensure that the applied loads do not exceed the proven allowable long-term load level to prevent creep deformations of the polymer materials, which would cause reduction in flow capacity.

8.0 MATERIALS

8.1 PIPE

When specified, pipe shall conform to the material and manufacturing requirements of the American Society for Testing and Materials (ASTM), American Association of State Highway and Transportation Officials (AASHTO), or American Water Works Association (AWWA) standards referenced. The ASTM standard and comparable AASHTO standard for a product are generally identical; however, there may be some differences, especially since AASHTO standards are normally a year behind ASTM standard revisions. Since these standards cover only material and manufacturing requirements, installation and performance requirements must be determined by the engineer and included in the construction contract documents. If there is a separate metric edition of a standard, its designation includes the letter M (i.e., C444M). The following lists of standards are for products commonly accepted and used on current projects; it is not intended that these listings restrict the use of new products that may be developed and found satisfactory.

8.1.1 Concrete Pipe

Reinforced and nonreinforced concrete pipe are used for gravity flow systems. Reinforced concrete pressure pipe and prestressed concrete pressure pipes are used for pressure flow systems. Concrete fittings and appurtenances such as wyes, tees, and manhole sections are generally available. A number of jointing methods are available depending on the tightness required and the operating pressure.

A number of mechanical processes are used in the manufacture of concrete pipe. These processes use various techniques, including centrifugation, vibration, packing, and tamping for consolidating the concrete in forms. Gravity and pressure concrete pipe may be manufactured to any reasonable strength requirement by varying the wall thickness, concrete strength, and quantity and configuration of reinforcing steel or prestressing elements.

Gravity Flow Applications

Concrete pipe is specified by nominal diameter, type of joint, and D-load strength or reinforcement requirements. The product should be manufactured in accordance with one or more of the following standard specifications:

ASTM C14/AASHTO M86 (ASTM C14M/ AASHTO M86M): Concrete Sewer, Storm Drain and Culvert Pipe—covers nonreinforced concrete pipe from 4- to 36-inch (100- to 900-mm) diameters in Class 1, 2, and 3 strengths.

ASTM C76/AASHTO M170 (ASTM C76M/ AASHTO M170M): Reinforced Concrete Culvert, Storm Drain, and Sewer Pipe—covers reinforced concrete pipe in five standard strengths. Class I—60- to 144-inch (1,500- to 3,600-mm) diameters; Class II, III, IV, and V—12- to 144-inch (300- to 3,600-mm) diameters. Larger sizes and higher classes are available as special designs.

ASTM C118 (ASTM C118M): Concrete Pipe for Irrigation or Drainage—covers concrete pipe intended to be used for the conveyance of water under low hydrostatic heads, generally not exceeding 25 feet (75 kPa) and for drainage in sizes from 4- to 24-inch (100- to 600-mm) diameters in standard and heavy-duty strengths.

ASTM C361 (ASTM C361M): Reinforced Concrete Low-Head Pressure Pipe—covers reinforced concrete pipe conduits with low internal hydrostatic heads generally not exceeding 125 feet (375 kPa) in sizes from 12- to 108-inch (100- to 2,700-mm) diameters.

ASTM C412/AASHTO M178 (ASTM C412M/AASHTO M178M): Concrete Drain Tile—covers non-reinforced concrete drain tile with internal diameters from 4- to 24-inch (100-to 600-mm) for standard quality, and to 36-inch (100- to 900-mm) for extra-quality, heavy-duty extra-quality, and special quality concrete drain tile.

ASTM C444/AASHTO M175 (ASTM C444M/AASHTO M175M): Perforated Concrete Pipe—covers perforated concrete pipe intended to be used for underdrainage in 4-inch (100-mm) and larger diameters.

ASTM C478/AASHTO M199 (ASTM C478M/AASHTO M199M): Precast Reinforced Concrete Manhole Sections—covers precast reinforced concrete manhole risers, grade rings, and tops to be used to construct manholes for storm and sanitary sewers.

ASTM C505 (ASTM C505M): Nonreinforced Concrete Irrigation Pipe with Rubber Gasket Joints—covers pipe to be used for the conveyance of water with working pressures, including hydraulic transients, of up to 30 feet (90 kPa) of head. Higher pressures may be used up to a maximum of 50 feet (150 kPa) for 6- to 12-inch (150- to 300-mm) diameters, and 40 feet (120 kPA) for 15- to 18-inch (375- to 450-mm) diameters by increasing the strength of the pipe.

ASTM C506/AASHTO M206 (ASTM C506M/AASHTO M206M): Reinforced Concrete Arch Culvert, Storm Drain, and Sewer Pipe—covers reinforced concrete arch pipe in sizes from 15- to 132-inch (375- to 3,300-mm) equivalent circular diameters. Larger sizes are available as special designs.

ASTM C507/AASHTO M207 (ASTM C507M/AASHTO M207M): Reinforced Concrete Elliptical Culvert, Storm Drain, and Sewer Pipe—covers reinforced elliptical concrete pipe in five standard classes of horizontal elliptical, 18- to 144-inch (450- to 3,600-mm) in equivalent circular diameter and five standard classes of vertical elliptical, 36 to 144 inches (900 to 3,600 mm) in equivalent circular diameter. Larger sizes are available as special designs.

ASTM C654/AASHTO M176 (ASTM C654M/AASHTO M176M): Porous Concrete Pipe—covers porous nonreinforced concrete pipe in sizes from 4- to 24-inch (100- to 600-mm) diameters and in two strength classes.

ASTM C655/AASHTO M242 (ASTM C655M/AASHTO M242M): Reinforced Concrete D-Load Culvert, Storm Drain and Sewer Pipe—covers accept-

ance of pipe design and production based on the D-load concept and statistical sampling techniques.

ASTM C789/AASHTO M259 (ASTM C789M/AASHTO M259M): Precast Reinforced Concrete Box Sections for Culverts, Storm Drains, and Sewers—covers box sections with 2 or more feet (0.6 m) of earth cover when subjected to highway live loads, and zero cover or greater when subjected to only dead loads in sizes from 3-foot (900-mm) span by 2-foot (600-mm) rise to 12-foot (3,600-mm) span by 12-foot (3,600-mm) rise.

ASTM C850/AASHTO M273 (ASTM C85O/AASHTO M273M): Precast Reinforced Concrete Box Sections for Culverts, Storm Drains, and Sewers with Less Than 2 Feet (0.6 m) of Cover Subject to Highway Loading—covers box sections with less than 2 feet (0.6 m) of earth cover in sizes from 3-foot (900-mm) span by 2-foot (600-mm) rise to 12-foot (3,600-mm) span by 12-foot (3,600-mm) rise.

ASTM C985 (ASTM C985): Nonreinforced Concrete Specified Strength Culvert, Storm Drain, and Sewer Pipe—covers acceptance of nonreinforced concrete pipe design and production based on specified strengths and statistical sampling techniques.

Pressure Flow Application

Concrete pressure pipe is specified to provide custom designs based on specific conditions of service. The product should be designed and manufactured in accordance with one or more of the following standards.

AWWA C300: Reinforced Concrete Cylinder Pipe—covers reinforced concrete cylinder pipe from 24- to 144-inch (600- to 3,600-mm) diameters and larger, and standard lengths in the 12- to 24-foot (3.6- to 7.2-m) range. Although the maximum loads and pressures for this type of pipe depend on the pipe diameter, wall thickness, and strength limitations of the concrete and steel, it is uncommon for this pipe to be used at pressures over 250 psi (1,750 kPa) or in trench installations exceeding 20 feet (6 m) of earth cover.

AWWA C301: Prestressed Concrete Cylinder Pipe—covers prestressed concrete cylinder pipe from 24- to 144-inch (600- to 3,600-mm) diameters and larger. Pipe larger than 250 inches (6,250 mm) in diameter has been produced. Standard lengths in the 16- to 24-foot range (4.8- to 7.2-m), and longer lengths are available. Prestressed concrete cylinder pipe has been designed for operating pressures over 400 psi (2,800 kPa) and earth covers in excess of 100 feet (30 m).

AWWA C302: Reinforced Concrete Noncylinder Pipe—covers reinforced concrete noncylinder pipe from 12- to 144-inch (300- to 3,600-mm) diameters and larger. Reinforced concrete noncylinder pipe, because it does not contain a watertight membrane (steel cylinder), is limited to internal pressures of 55 psi (385 kPa) or less.

AWWA C303: Pretensioned Concrete Cylinder Pipe—covers pretensioned concrete cylinder pipe from 10- to 48-inch (250- to 1,200-mm) diameters and larger. Standard lengths are generally 36 to 40 feet (10.8 to 12 m). Pretensioned concrete cylinder pipe has been designed for pressures over 400 psi (2,800 kPa).

8.1.2 Thermoplastic Pipe

Thermoplastic pipe materials include a broad variety of plastics that can be repeatedly softened by heating and hardened by cooling through a temperature range characteristic for each specific plastic, and in the softened state can be shaped by molding or extrusion. Generally, thermoplastic pipe materials are limited to acrylonitrile-butadiene-styrene (ABS), polyethylene (PE), and polyvinyl chloride (PVC). Thermoplastic pipes are produced in a variety of shapes and dimensions. Pipe properties can be modified by changing the wall thickness or profile, for both pressure and nonpressure applications.

8.1.2.1 Acrylonitrile-Butadiene-Styrene (ABS) Pipe

Acrylonitrile-Butadiene-Styrene (ABS) pipe is manufactured by extrusion of ABS material and is limited to gravity flow applications. ABS composite pipe is manufactured by extrusion of ABS material with a series of truss annuli that are filled with filler material such as lightweight Portland cement concrete. ABS fittings are available for the product. The jointing systems available include elastomeric gasket joints and solvent cement joints.

Gravity Flow Applications

ABS pipe should be manufactured in accordance with one of the following standard specifications.

ASTM D2680/AASHTO M264: Acrylonitrile-Butadiene-Styrene (ABS) and Poly (Vinyl Chloride) (PVC) Composite Sewer Piping—covers ABS or PVC composite pipe, fittings, and a joining system for nonpressure systems.

ASTM D2751: Acrylonitrile-Butadiene-Styrene (ABS) Sewer Pipe and Fittings—covers ABS pipe and fittings from 3- to 12-inch (75- to 300-mm) diameter.

8.1.2.2 Polyethylene (PE) Pipe

PE pipe is used for both gravity and pressure flow systems. PE pipe is manufactured by extrusion of PE plastic material. PE pipe is specified by material designation, nominal diameter (inside or outside), standard dimension ratios, ring stiffness, and type of joint. PE fittings are available.

Gravity Flow Applications

PE pipe for gravity flow applications should be manufactured in accordance with one or more of the following standard specifications.

ASTM F405/AASHTO M252: Corrugated Polyethylene Tubing and Fittings—ASTM F405 covers pipe from 3- to 6-inch (75- to 150-mm) diameters. AASHTO M252 covers pipe from 3- to 10-inch (75- to 250-mm) diameters.

ASTM F667: Large Diameter Corrugated Polyethylene Tubing and Fittings—ASTM F667 covers pipe from 8- to 24-inch (200- to 600-mm) diameters. AASHTO M294 covers pipe from 12- to 36-inch (300- to 900-mm) diameters.

AASHTO M294 Corrugated Polyethylene Pipe (12-inch to 36-inch diameter pipe).

ASTM F810: Smoothwall Polyethylene (PE) Pipe for Use in Drainage and Waste Disposal Absorption Fields—covers smoothwall PE pipe, including coextruded, perforated, and nonperforated from 3- to 6-inch (75 to 150-mm) diameters.

ASTM F892: Polyethylene (PE) Corrugated Pipe with a Smooth Interior and Fittings—covers corrugated PE pipe 4 inches (100 mm) in diameter.

ASTM F894: Polyethylene (PE) Large Diameter Profile Wall Sewer and Drain Pipe—covers profile wall PE pipe from 18- to 120-inch (450- to 3,000-mm) diameters for both low-pressure and gravity flow applications.

Pressure Flow Applications

PE pressure pipe should be manufactured in accordance with one or more of the following standard specifications.

ASTM D2239: Polyethylene (PE) Plastic Pipe (SIDR-PR) Based on Controlled Inside Diameter—covers PE pipe with 0.5- to 6-inch (12- to 150-mm) diameters and with pressure ratings from 400 to 800 psi (2,800 to 5,600 kPa).

ASTM D3035: Polyethylene (PE) Plastic Pipe (SDR-PR) Based on Controlled Outside Diameter—covers PE pipe 0.5- to 6-inch (12- to 150-mm) diameters and with pressure ratings from 400 to 800 psi (2,800 to 5,600 kPa).

8.1.2.3 Polyvinyl Chloride (PVC) Pipe

PVC pipe is used for both gravity and pressure flow systems. PVC pipe is manufactured by extrusion of the material. PVC composite pipe is manufactured by extrusion of PVC material with a series of truss annuli that are filled with material such as lightweight

Portland cement concrete. PVC pipe is specified by nominal diameter, dimension ratio, pipe stiffness, and type of joint. PVC pressure and nonpressure fittings are available.

Gravity Flow Applications

PVC pipe for gravity flow applications should be manufactured in accordance with one or more of the following standard specifications.

ASTM D2680/AASHTO M264: Acrylonitrile-Butadiene-Styrene (ABS) and Poly (Vinyl Chloride) (PVC) Composite Sewer Piping—covers ABS or PVC composite pipe, fittings, and a joining system for non-pressure sanitary sewer and storm drain systems in 6- to 15-inch (150- to 375-mm) diameters.

ASTM D2729: Poly (Vinyl Chloride) (PVC) Sewer Pipe and Fittings—covers material and test requirements for PVC pipe and fittings for sewer and drain pipe in sizes from 2- to 6-inch (50- to 150-mm) diameters. Standard perforations available only in 4-inch (100-mm) diameter pipe.

ASTM D3034: Type PSM Poly (Vinyl Chloride) (PVC) Sewer Pipe and Fittings—covers material and test requirements for PVC pipe and fittings for sewer pipe systems from 4- to 15-inch (100- to 375-mm) diameters.

ASTM F679: Poly (Vinyl Chloride) (PVC) Large Diameter Plastic Gravity Sewer Pipe and Fittings—covers material and test requirements for PVC gravity sewer pipe and fittings from 18- to 27-inch (450- to 675-mm) diameters, with integral bell elastomeric seal joints and smooth inner walls.

ASTM F758: Smooth-Wall Poly (Vinyl Chloride) (PVC) Plastic Underdrain Systems for Highways, Airport, and Similar Drainage—covers material and test requirements for smooth wall pipe and fittings for PVC underdrains from 4- to 8-inch (100- to 200-mm) diameters with perforated or nonperforated walls for use in subsurface drainage systems.

ASTM F789: Type PS-46 Poly (Vinyl Chloride) (PVC) Plastic Gravity Flow Sewer Pipe and Fittings—covers materials and test requirements for PVC gravity sewer pipe and fittings from 4- to 18-inch (100- to 450-mm) diameters, with a minimum pipe stiffness of 46 psi (320 kPa).

ASTM F794: Poly (Vinyl Chloride) (PVC) Large Diameter Ribbed Gravity Sewer Pipe and Fittings Based on Controlled Inside Diameter—covers materials and test requirements for PVC gravity sewer ribbed pipe and fittings from 4- to 48-inch (200- to 1,200-mm) inside diameters, with integral bell and elastomeric seal joints.

ASTM F800: Corrugated Poly (Vinyl Chloride) (PVC) Tubing and Compatible Fittings—covers mate-rials and test requirements for perforated and nonper-forated PVC tubing and fittings from 4- to 12-inch (100- to 300-mm) diameters for use in agricultural and other soil drainage and septic tank effluent beds.

ASTM F949: Poly (Vinyl Chloride) (PVC) Corru-gated Sewer Pipe with a Smooth Interior and Fittings—covers materials and test requirements for PVC pipe and fittings from 4- to 10-inch (100- to 250-mm) diam-eters with corrugated outer wall fused to a smooth inner wall for sanitary and storm sewers and perforated and nonperforated pipe for subdrainage.

Pressure Flow Applications

PVC pressure pipe should be manufactured in accor-dance with one of the following standard specifications.

ASTM D1785: Poly (Vinyl Chloride) (PVC) Plas-tic Pipe, Schedules 40, 80, 120—covers materials and test requirements for PVC pipe pressure rated for water transmission in diameters ranging from 1/8- to 12-inch (3- to 300-mm) diameters.

ASTM D2241: Poly (Vinyl Chloride) (PVC) Pres-sure Rated Pipe (SDR Series)—covers materials and test requirements for PVC pipe pressure rated for water transmission in sizes ranging from 1/8- to 36-inch (3- to 900-mm) outside diameters. Available with SDRs ranging from 13.5 to 64, depending on the diameter.

AWWA C900: Poly (Vinyl Chloride) (PVC) Pres-sure Pipe, 4 in. (100-mm) through 12 in. (300-mm), for Water—covers materials and test requirements for PVC pipe pressure rated for water transmission in sizes ranging from 4- to 12-inch (100- to 300-mm) diameters, and three pressure classes of 100, 150, and 200 psi (700, 1,050, and 1,400 kPa) with DRs of 25, 18, and 14, respectively, and providing a hydrostatic design basis (HDB) of 4,000 psi (28 MPa).

AWWA C905: Poly (Vinyl Chloride) (PVC) Water Transmission Pipe, Nominal Diameters 14 in. (350 mm) through 36 in. (900 mm)—covers materials and test requirements for PVC pipe pressure rated for water transmission in sizes ranging from 14- to 36-inch (350- to 900-mm) outside diameters, and DRs of 18, 21, 25, 26, 32.5, and 41, and providing a hydro-static design basis (HDB) of 4,000 psi (28 MPa) for pressure ratings ranging from 100 to 235 psi (700 to 1650 kPa).

8.1.3 Metal Pipe

Corrugated metal pipe is fabricated from corru-gated steel or aluminum sheets or coils. Corrugated metal pipe is specified by size, shape, wall profile, gauge or wall thickness, and coating or lining. Appurtenances, including tees, wyes, elbows, and manholes, are available. Corrugated metal pipe is lim-ited to gravity flow applications.

Gravity Flow Applications

Corrugated metal pipe should be manufactured in accordance with one or more of the following standard specifications.

ASTM A760/AASHTO M36: Corrugated Steel Pipe, Metallic-Coated for Sewers and Drains—covers metallic-coated corrugated steel pipe from 4- to 144-inch (100- to 3,600-mm) diameters.

ASTM A762/AASHTO M245: Corrugated Steel Pipe, Polymer Precoated for Sewers and Drains—covers polymer precoated corrugated steel pipe from 4- to 144-inch (100- to 3600-mm) diameters.

ASTM A849/AASHTO M190: Post Coated (Bituminous) Corrugated Steel Sewer and Drainage Pipe—covers post coated corrugated steel pipe from 4- to 144-inch (100- to 3,600-mm) diameters.

ASTM B745/AASHTO M196: Corrugated Aluminum Pipe for Sewers and Drains—covers corrugated aluminum pipe from 4- to 144-inch (100- to 3,600-mm) diameters.

8.1.4 Vitrified Clay Pipe (VCP)

VCP is manufactured from clays and shales and vitrified at high temperatures. VCP is available in standard and extra-strength classifications, and is specified by nominal pipe diameter, strength, and type of joint. The product is limited to gravity flow applications.

Gravity Flow Applications

The product should be manufactured in accordance with one or more of the following standard specifications.

ASTM C4/AASHTO M179: Clay Drain Tile—covers drain tile from 4- to 30-inch (100- to 750-mm) diameters in standard, extra-quality, and heavy-duty strengths.

ASTM C498: Clay Drain Tile, Perforated—covers perforated drain tile from 4- to 18-inch (100- to 450-mm) diameters in strengths of standard, extra-quality, heavy-duty, and extra-strength.

ASTM C700/M65: Clay Pipe, Vitrified, Extra Strength, Standard Strength, and Perforated—covers perforated and nonperforated pipe from 3- to 42-inch (75- to 1,050-mm) diameters in extra-strength and standard strength.

8.2 OTHER MATERIALS AND PRODUCTS

Geocomposites, geomembranes, geotextiles, aggregates, wick drains, and pump and lift stations are not covered by national standard specifications. The requirements for such materials and products must be specified in construction contract documents by the engineer.

9.0 CODES AND PERMITS

9.1 GENERAL

In the concept stages of an urban subsurface drainage project, preferably before the site analysis and system configuration phases, it is important to obtain copies and an understanding of all applicable federal, state, and local codes. At the same time, all federal, state, and local permits for the project should be identified, and the requirements and submittal timing of each clearly understood.

9.2 CODES

Federal, state, and local codes that apply to the design, construction, and operation of an urban subsurface drainage system shall be considered.

9.3 PERMITS

Federal, state, and local temporary permits that are necessary prior to and during construction of an urban subsurface drainage system project shall be secured by the owner, owner's agent, or contractor prior to construction. Any permanent permits, such as the Corps of Engineers' 404 Permit, which must be maintained after construction of the project, shall be secured by the owner and/or the owner's agent.

A copy of any temporary and permanent permits secured by the owner or owner's agent, shall be included as part of the contract documents. Copies of all permits secured by the owner's agent or the contractor should be furnished to the owner.

The contractor is responsible to conform to the terms and provisions of all permits during construction as stated in the contract documents.

10.0 REFERENCES

10.1 REFERENCE DOCUMENTS

American Association of State Highway and Transportation Officials, *Standard Specifications for Highway Bridges*, Washington, D.C., 1992.

American Concrete Pipe Association, *Concrete Pipe Design Manual*, ACPA, Vienna, Va., 1992.

ACPA, *Concrete Pipe Handbook*, ACPA, Vienna, Va., 1988.

American Iron and Steel Institute, *Handbook of Steel Drainage and Highway Construction Products*, AISI, New York, 1971.

American Railway Engineering Association, *Manual for Railway Engineering*, AREA, Washington, D.C., 1993.

American Society of Civil Engineers, *Nomenclature for Hydraulics,* ASCE Manuals and Reports of Practice No. 43, ASCE, New York, 1962.

American Society for Testing and Materials, *Standard Test Method for Constant Head Hydraulic Transmissitivity (In-Plane Flow) of Geotextiles and Geotextile Related Products*, ASTM D4716, ASTM, Philadelphia, 1992.

ASTM, *Standard Test Method for Permeability of Granular Soils (Constant Head)*, AASHTO T215, E1-1993 R, ASTM D2434, ASTM, Philadelphia, 1994.

ASTM, *Standard Practice for Underground Installation of Thermoplastic Pipe for Sewers and Other Gravity-Flow Applications*, ASTM D2321-00, ASTM, Philadelphia, 2000.

Chambers, R. E., McGrath, T. J., and Heger, F. J., *Plastic Pipe for Subsurface Drainage of Transportation Facilities*, Transportation Research Board, National Cooperative Highway Research Program Report 225, Washington, D.C., October1980.

Federal Aviation Administration, U.S. Department of Transportation, *Airport Drainage*, ACI 50/5230-5B, Washington, D.C., 1970.

Federal Highway Administration, *Durable Pavement Systems, Participants' Notebook*, Office of Technology Applications and Office of Engineering, Washington, D.C., 1992.

Graber, S. D. Collection Conduits Including Subsurface Drains. *Journal of Environmental Engineering*, Vol. 130, No. 1, American Society of Civil Engineers, pp. 67–80, January 2004.

NRCS, SCS National Engineering, *Field Handbook*, Subchapter C. Sect. 650, 1428 (b), USDA, NRCS, Washington, D.C., 20250-0016, 1991.

National Clay Pipe Institute, *Clay Pipe Engineering Manual*, NCPI, Washington, D.C., 1982.

Sherard, J. L., Dunnigan, L. P., and Talbot., J. R., Basic Properties of Sand and Gravel Filters, *Journal of Geotechnical Engineering*, Vol. 110, No. 6, American Society of Civil Engineers, pp. 684–700, June 1984a.

Sherard, J. L., Dunnigan, L. P., and Talbot, J. R., Filters for Silts and Clays. *Journal of Geotechnical Engineering*, Vol. 110, No. 6, American Society of Civil Engineers, pp. 701–718, June 1984b.

Spangler, M. G., *Soil Engineering,* International Textbook Co., Scranton, Penn., 1966.

TeKrony, R. G., Sanders, G. D., and Cummins, B., History of Drainage in the Bureau of Reclamation, *Journal of Irrigation & Drainage Engineering*, Vol. 130, No. 2, American Society of Civil Engineers, pp. 148–153, March/April 2004.

U.S. Army Eng. Waterways Exp. Sta., Corps of Engineers, *Investigation of Filter Requirements for Underdrains*, Tech Memo. No. 183-1, 35 p., 1941.

U.S. Department of Agriculture, Soil Conservation Service, Structural Design, Section 6, *National Engineering Handbook,* USDA, Washington, D.C., December 1980.

U.S. Department of Agriculture, Soil Conservation Service, *The Structural Design of Underground Conduits,* Technical Release No. 5, USDA, Washington, D.C., November 1958.

U.S. Department of Agriculture, Soil Conservation Service, Drainage of Agricultural Land, Section 16, *National Engineering Handbook*, USDA, Washington, D.C., 1971.

U.S. Department of the Interior, Bureau of Reclamation, *Drainage Manual*, USDI, Denver, Colo., 1993.

U.S. Department of the Interior, Bureau of Reclamation, *Water Measurement Manual*, USDI, 1997.

Uni-Bell PVC Pipe Association, *Handbook of PVC Pipe: Design and Construction*, Dallas, 1986.

Wyatt, T., Baker, W., and Hall, J., *Drainage Requirements in Pavements—DRIP*, FHWA-SA-96-070, January 1998.

10.2 GENERAL REFERENCES

Advanced Drainage Systems, Inc., *Specifier Manual*, Columbus, Ohio, 1984.

American Concrete Pipe Association, *Concrete Pipe Installation Manual*, ACPA, Vienna, Va., 1988.

American Society of Agricultural Engineers, *Hydrologic Modeling of Small Watersheds*, ASAE, St. Joseph, Mich., 1982.

American Society for Testing and Materials, *Standard Test Method for Compressive Properties of Rigid Cellular Plastics*, ASTM D1621, ASTM, Philadelphia, 1992.

ASTM, *Standard Practice for Underground Installation of Thermoplastic Pipe for Sewers and Other Gravity-Flow Applications,* ASTM D2321-00, ASTM, 2000.

American Society of Civil Engineers, *Gravity Sanitary Sewer Design and Construction*, ASCE

Manuals and Reports on Engineering Practice No. 60, ASCE, New York, 1982.

ASCE, "In-Plane Composite Drains," *Civil Engineering,* ASCE, New York, pp. 48–51, August 1984.

ASCE, *Urban Subsurface Drainage,* ASCE Manuals and Reports on Engineering Practice No. 95, ASCE, New York, 1998.

Anderson, B., *Underground Waterproofing,* WEBCO, Stillwater, Minn., 1983.

Beck, D. E., Testing and Comparing Geocomposite Drainage Products, *Geotechnical Fabrics Report*, Industrial Fabrics Association International, St. Paul, Minn., July/August 1988.

Bouwer, H., *Groundwater Hydrology,* McGraw-Hill, New York, 1978.

Cedergren, H. R., *Seepage, Drainage and Flow Nets*, John Wiley & Sons, New York, 1967.

Cedergren, H. R., *Drainage of Highways and Airfield Pavements,* John Wiley & Sons, New York, 1974.

Chamber's Technical Dictionary, 3rd Edition, Macmillan Co., New York, 1967.

Davis, C. V. and Sorensen, K. E., *Handbook of Applied Hydraulics*, McGraw-Hill, New York, 1986.

Davis, S. N. and DeWiest, R. J. M., *Hydrogeology*, John Wiley & Sons, New York, 1966.

Dempsey, B. J., *Pavement Drainage System Design*, prepared for Wisconsin DOT, February 15–16, 1988.

Driscoll, F. G., Ed., *Ground Water and Wells,* 2nd Edition, Johnson Division UOP, St. Paul, Minn., 1986.

Engineers Joint Council, *Thesaurus of Engineering and Scientific Terms*, New York, December 1967.

Federal Highway Administration, *Evaluation of Test Methods and Use Criteria for Geotechnical Fabrics in Highway Applications*, Report No. FHWA/RD-80-021, Washington, D.C., 1980.

FHA, *Design of Urban Highway Drainage*, Report No. FHWA-TS-79-225, Washington, D.C., 1983.

FHA, *Highway Subdrainage Design*, Report No. FHWA-TS-224, Washington, D.C., 1980.

FHA, *Hydraulic Design of Highway Culverts*, Hydraulic Design Series No. 5, Washington, D.C., 1985.

FHA, *Design of Highway Drainage—The State of the Art,* Report No. FHWA-TS-79-225, Washington, D.C., 1979.

Fetter, C. W., Jr., *Applied Hydrogeology,* Charles E. Merrill, Columbus, Ohio, 1980.

Freeze, R. A. and Cherry, J. A., *Groundwater,* Prentice-Hall, Englewood Cliffs, N.J., 1979.

Geosystems, Inc., Vertical Drains, *GeoNotes—A Ground Improvements Update,* Sterling, Va., undated.

Hancor, Inc., *Recommended Installation Practice for Hancor Hi-Q Titelines, Heavy Duty and Heavy Duty-AASHTO Pipe*, Findlay, Ohio, 1993.

Hannon, J. D. and California DOT, *Underground Disposal of Storm Water Runoff, Design Guidelines Manual*, FHWA-TS-80-218, Federal Highway Administration, Washington, D.C., 1980.

Hem, J. D., *Study and Interpretation of the Chemical Characteristics of Natural Water,* U.S. Geological Survey Water-Supply Paper 1473, Washington, D.C., 1970.

Illinois Department of Transportation, *Highway Standards Manual*, Springfield, Ill., November 1993.

Koerner, R. M., *Designing with Geosynthetics,* 2nd Edition, Prentice-Hall, Englewood Cliffs, N.J., 1990.

Lafayette Farm & Industry, *Agri-Fabric Awareness Manual,* Cuba City, Wis., undated.

Lohman, S. W. et al., *Definitions of Selected Ground Water Terms-Revisions and Conceptual Refinements*, Geological Survey Water-Supply Paper 1988, U.S. Geological Survey, Washington, D.C., 1972.

Merritt, F. S., *Standard Handbook for Civil Engineers*, McGraw-Hill, New York, 1983.

Peck, R. B., Hanson, W. E., and Thornburn, T. H., *Foundation Engineering,* John Wiley & Sons, New York, 1974.

Powers, J. P., *Construction Dewatering,* John Wiley & Sons, New York, 1979.

Roister, D. L., *Landslide Remedial Measures*, Tennessee Department of Transportation, Nashville, Tenn., 1982.

Sacks, A. M., "Geosynthetics," p. 14, *Remodeling Magazine,* Hanley Woods Inc., Washington, D.C., November 1987.

Sacks, A., "R_x for Basement Water Problems," *The Family Handyman,* St. Paul, Minn., pp. 36–40, September 1981.

Schuster, R. L. and Kruse, R. J., eds. *Landslides Analysis and Control, Transportation Research Board*, Special Report 176, Washington, D.C., 1978.

Schwab, G. O., Revert, R. K., et al., *Soil and Water Conservation Engineering,* 3rd Edition, John Wiley & Sons, New York, 1981.

Sowers, G. F., *Introductory Soil Mechanics and Foundations: Geotechnical Engineering,* Macmillan Publishing Co., New York, 1979.

Todd, D. K., *Ground Water Hydrology,* John Wiley & Sons, New York, 1980.

U.S. Department of Agriculture, "Soil Conservation Service, Drainage," Chapter 14 in: *Engineering Field Manual,* USDA, Washington, D.C., 1969.

USDA, *Soil Conservation Service, Technical Guide*, Section IV, Standard 606, Subsurface Drainage, May 1988.

U.S. Department of the Interior, Bureau of Reclamation, Ground Water Manual, Washington, D.C., 1995.

Vetch, J. O. and Humphrey, C. Rs, *Water and Water Use Terminology,* Thomas Printing & Publishing Co., New York, 1966.

Standard Guidelines for the Installation of Urban Subsurface Drainage

CONTENTS

FOREWORD

The *Standard Guidelines for the Installation of Urban Subsurface Drainage* is an independent document intended to complement the ASCE Manuals and Reports on Engineering Practice No. 95, *Urban Subsurface Drainage*. These standard guidelines are companions to the *Standard Guidelines for the Design of Urban Subsurface Drainage* and *Standard Guidelines for the Operation and Maintenance of Urban Subsurface Drainage*. These standard guidelines were developed by the Urban Drainage Standards Committee, which is responsible to the Environmental and Water Resources Institute of the American Society of Civil Engineers.

The material presented in this publication has been prepared in accordance with recognized engineering principles. These standard guidelines should be used only under the direction of an engineer who is competent in the field of urban subsurface drainage. The publication of the material contained herein is not intended as a representation or warranty on the part of the American Society of Civil Engineers, or of any other person named herein, that this information is suitable for any general or particular use, or promises freedom from infringement of any patent or patents. Anyone making use of this information assumes all liability from such use.

ACKNOWLEDGEMENTS

The American Society of Civil Engineers (ASCE) acknowledges the work of the Urban Drainage Standards Committee of the Environmental and Water Resources Institute of ASCE (EWRI of ASCE).

This group comprises individuals from many backgrounds, including consulting engineering, research, the construction industry, education, and government. Those individuals who serve on the Urban Drainage Standards Committee are:

William Curtis Archdeacon, Chair
Richard H. Berich
Christopher B. Burke
Robert T. Chuck
F. Scott Dull
Robert S. Giurato, Secretary
S. David Graber
Jay M. Herskowitz
Conrad G. Keyes, Jr.
John M. Kurdziel
John J. Meyer
Philip M. Meyer
James R. Noll

Walter J. Ochs
Garvin J. Pederson
Glen D. Sanders
Erez Sela
Alan N. Sirkin
Edward L. Tharp
William J. Weaver
Richard D. Wenberg, Past Chair
David L. Westerling
Stan E. Wildesen
Lyman S. Willardson (deceased)
Donald E. Woodward

Standard Guidelines for the Installation of Urban Subsurface Drainage

1.0 SCOPE

The intent of this standard is to present installation and construction guidance for urban subsurface drainage systems. The collection and conveyance of subsurface drainage waters are within the purview of this standard for applications such as airports, roads, and other commercial transportation systems, and industrial, residential, and recreational areas. Incidental surface water is considered.

This standard does not address agricultural drainage, landfills, recharge systems, detention ponds, conventional storm sewer systems, or the use of injection systems.

Customary units and standard international (SI) units are used throughout this document.

2.0 DEFINITIONS

2.1 GENERAL

This section defines specific terms for use in these guidelines. References included in Section 3.0 may also be helpful in understanding the terms of these guidelines.

2.2 TERMS

Bedding—Granular material placed around subsurface drains to provide structural support for the drain.

Drain envelope—Generic name for materials placed on or around a drainage product, irrespective of whether used for mechanical support (bedding), hydraulic purposes (hydraulic envelope), or to stabilize surrounding soil material (filter envelope).

Filter envelope—Permeable material placed around a drainage product to stabilize the structure of the surrounding soil material. A filter envelope may initially allow some fines and colloidal material to pass through it and into the drain.

Geocomposite—Geosynthetic materials for collecting and transporting water while maintaining soil stability.

Geomembrane—Sheet materials intended to form an impervious barrier.

Geosynthetic—Synthetic material or structure used as an integral part of a project, structure, or system.

Within this category are subsurface drainage and water control materials such as geomembranes, geotextiles, and geocomposites.

Geotextile—Woven or nonwoven engineering fabric sheet material intended to allow the passage of water (but not fines, and without collecting fines at the soil–textile interface).

Grade—May refer to either (1) the slope of the drain in vertical units/horizontal units, or (2) the specified vertical location of the drain, depending on the context in which it is used.

Hydraulic envelope—Permeable material placed around a drainage product to improve flow conditions in the area immediately adjacent to the drain.

OSHA—Occupational Safety and Health Administration, the federal agency responsible for safety and health concerns on construction job sites.

Subsurface water—All water beneath the ground or pavement surface. Sometimes referred to as groundwater.

3.0 CONTRACT DOCUMENTS

3.1 GENERAL

Plans and specifications typically make up the construction contract documents. Other documents become part of the contract by reference.

3.2 PLANS

Plans are drawings, most prepared to a scale, showing the proposed subsurface drainage system and known surface and subsurface features that may affect the new installation and construction. The plans typically show type, size, material, grade, and location of the subsurface drainage system.

3.3 SPECIFICATIONS

Specifications are written text and/or details that provide specifics of the construction. They typically detail or reference all appropriate standards required for the product and project. Construction methods may be specified on a case-by-case basis.

3.4 OTHER

Many other publications are made part of the contract documents by reference only, such as government requirements, permits, reports, or trade and industry specifications. Some of these reports will include soil borings, past construction observations, and studies.

4.0 SITE INVESTIGATION

4.1 GENERAL

It is necessary to examine the plans and specifications and make a personal examination of the site and its surroundings prior to construction. This investigation should include reviews of both surface features and subsurface features.

4.2 SURFACE FEATURES

The surface features of the site should be located through a topographic survey and shown on the plans. The plans should be compared with existing field conditions to determine whether there are any differences between the topographic survey and present conditions. Discrepancies are to be brought to the attention of the engineer or project manager.

4.3 SUBSURFACE FEATURES

Subsurface features principally consist of utilities, water table level, and geologic conditions. All subsurface features affecting the work should be determined and shown on the plans. If called for in the plans, the contractor shall undertake a program of field surveys and test pits to verify subsurface conditions at specific locations identified in the plans. The contractor shall immediately notify the engineer if changed conditions are found and shall not commence construction of the work in areas where discrepancies are noted until the engineer has issued written clarification.

4.3.1 Utilities

The location and size of sanitary sewers, drains, culverts, gas lines, water mains, electric lines, telephone conduits, and other underground utilities and structures should be shown on the plans. This information should be obtained from both field surveys and other available records. Utility locating organizations are frequently available in urban areas to mark the position of lines. The level of confidence for utility location and depth may require special provisions to be placed in the contract documents, indicating areas where contractors must physically locate existing utilities before constructing drain lines.

4.3.2 Geologic Conditions

All appropriate and available geologic conditions should be shown on the plans. An assessment should be made with respect to rock and groundwater conditions.

5.0 INSTALLATION

5.1 GENERAL

Prior to construction, all documents, including plans and specifications, subsurface information, standard details, product shop drawings, and special provisions should be reviewed and any questions resolved.

5.2 SAFETY

The contractor is responsible for construction site safety. Federal regulations covering safety for all types of construction are published in the Safety and Health Regulations for Construction under the Department of Labor, Occupational Safety and Health Administration (OSHA). Many states, municipalities, and other local agencies have established codes of safe practice regarding construction. These regulations apply to all types of construction, including alteration and repair work. All personnel associated with the construction should be familiar with the requirements applicable to subsurface drainage system projects, especially in regard to safe trenching procedures.

5.3 SOIL EROSION AND SEDIMENT CONTROL

Erosion and sediment control at the site should be in accordance with federal, state, municipal, and local agency regulations, and as otherwise established by the contract plans and specifications.

5.4 SITE PREPARATION

Site preparation should be in accordance with the contract plans and specifications and may include topsoil stripping, clearing and grubbing, pavement and sidewalk removal, rough grading, protection or relocation of existing natural drainage, removal of unsuitable soil material, construction of access roads, detours, and protection or relocation of existing structures and utilities.

5.5 MATERIALS RECEIVING, HANDLING, AND STORAGE

The contractor is responsible for receiving, proper handling, and storage of all construction materials for the project. Materials damaged in shipment or at the site that cannot be repaired should be marked or tagged and removed from the site. All materials should be unloaded and handled with reasonable care. Stockpiling of materials should be as near as possible to where they will be installed, consistent with municipal requirements, safety, and environmental considerations. All materials should be stored as recommended by the manufacturer. All materials should be stockpiled in a safe manner.

5.6 LINE AND GRADE

The contractor is responsible for maintaining all line and grade, monuments, control points, and stakes set by the engineer or surveyor until the project is completed and accepted.

For all subsurface drainage systems, line and grade should be in accordance with the contract plans and specifications.

All subsurface drainage systems should be installed true to line and grade in accordance with the contract plans and specifications. Adjustments to correct departures from specified line and grade shall not exceed those permitted by contract documents or manufacturers' recommendations, whichever are more restrictive, provided that such corrections never result in a reversal of the slope in the drainage system. Moreover, realignments must never result in buckling or other deformations of flexible drainage materials, such as flexible pipes and geocomposites that damage or reduce the flow capacity of the system. The return to specified line and grade should be made by adding or removing bedding material, and the use of wedges or blocks is unacceptable.

The maximum allowable departure from the horizontal alignment should be specified by the contract documents. Departure distances must be determined by measurement along common elements of the planned versus actual installation, such as centerline-to-centerline measurements.

The relatively complex hydraulic characteristics of geocomposite drainage systems require that strict controls be placed on departures from specified vertical alignment (grade). When departures occur, the return to specified grade should be at a rate not exceeding 15% of the specified grade or at a rate established by the specifications.

5.7 EXCAVATION

Excavation should be in accordance with contract plans and specifications and may include trenching, backfilling, embankment construction, soil stabilization, and control of groundwater and surface drainage. Adequate knowledge of subsurface conditions is required for all types of excavation. Additional exploration and analysis are recommended if the subsurface information on the plans is insufficient. The contractor should notify the owner if archaeological items are encountered during construction.

5.7.1 Excavation Limits
Excavation, installation, and backfill operations should be performed in a timely manner to reduce open trench time. The length of open trench should comply with limits established by OSHA or applicable state and local regulations. Specified trench width requirements for flexible conduits should also be maintained to ensure proper deflection control. Trench depth and width should be in accordance with the contract plans and specifications. If the trench width becomes greater than specified, the contractor should contact the engineer for a reevaluation of the required pipe strength, bedding materials and methods, and backfilling procedures to be used.

5.7.2 Handling of Excavated Material
Excavated material to be used as backfill should be stockpiled in accordance with the contract plans and specifications and applicable safety regulations a safe distance back from the edge of the trench. Generally, if trench walls are unsupported, the minimum distance from the trench side to the excavated material should be either 3 feet (1 m) or one-half of the trench depth, whichever is greater. If the trench walls are supported, the usual minimum distance from the trench side to the excavated material should be 3 feet (1 m).

These general recommendations should not supersede job-specific requirements in the presence of unstable soils and/or the potential for accumulation of water in the trenches.

For trench installations where drainage material and backfill are placed simultaneously and where personnel are not allowed in the trench, soil placement may be unrestricted. Unused excavated materials should be disposed of in accordance with contract documents.

5.7.3 Sheathing and Shoring

OSHA and many states, municipalities, and other local agencies have established codes of safe practice regarding support requirements for trench excavation. Sheathing and shoring must be adequate to prevent cave-in of the trench walls or subsidence of areas adjacent to the trench and to prevent sloughing of the base of the excavation. Sheathing and shoring should not extend into the soil envelope zone of the pipe or geocomposite drainage system. Any sheathing placed below the top of the drainage product of flexible pipes or geocomposite drainage materials or the springline of rigid pipes should remain in place after backfilling. Movement of shoring following backfill placement may reduce the structural integrity of the surrounding embedment material.

The contractor is responsible for adequacy of any required sheathing and shoring. The strength of support systems should be based on the principles of geotechnical and structural engineering as applicable to the materials encountered. Refer to the appendix for recommended use of trench boxes.

5.7.4 Dewatering

When necessary, all excavations should be dewatered prior to and during installation and backfilling of the subsurface drain system. The contractor is responsible for dewatering operations and should ensure that foundation and bedding materials are not being removed through the dewatering system and that property damage does not result.

The discharge from dewatering systems should be released in such a way that it does not harm existing structures or the environment and that it does not, by reinfiltration, put an additional load on the dewatering system.

5.8 FOUNDATION PREPARATION

Bedding for subsurface drainage systems should be as specified and completed to design line and grade. The intrusion of foreign material into any portion of the drainage system due to construction and weather events should be prevented until the system is adequately protected by backfill.

5.9 PLACEMENT OF DRAINAGE MATERIALS

The pipe materials covered are pipe, geocomposites, and other.

5.9.1 Pipe

A bedding material should be placed on the foundation, the pipe laid and connected, and backfill placed, all in accordance with the contract plans and specifications.

5.9.2 Geocomposite Drainage Materials

Prefabricated geocomposite subsurface collector drains may be placed in trenches by hand or machine. Installation techniques must not cause damage to the interior core and geotextile overwrap, including any factory-made seams and connections. Joints should be made using connectors recommended by the manufacturer or in accordance with the contract plans and specifications. All joints should be made soil-tight using tape, glue, or other sealing procedures recommended by the manufacturer or in accordance with the contract plans and specifications. In all cases, such sealing procedures should provide ensured long-term resistance to degradation in wet subsurface environments.

5.9.3 Other Drainage Materials

Geomembranes, geotextiles, aggregates, and pump and lift stations should be installed in accordance with the contract plans and specifications.

5.10 BACKFILL

Backfill material should be placed and compacted in accordance with the contract plans and specifications. Backfill should not be dumped or dropped directly on any portion of the drainage system. Heavy equipment operations should be controlled so as not to damage any portion of the drainage system. Backfill material should be placed in layers in accordance with contract documents and should be compacted.

5.11 SITE RESTORATION

Restoration of grass, shrubs, and other plantings should be performed in accordance with contract documents. Until revegetation is complete, adequate protection against erosion and runoff is necessary and should be in accordance with the contract documents and governing regulations. All revegetation and tree repair should be in accordance with accepted horticultural practice.

When replacing permanent pavement, the subgrade must be restored and compacted until smooth and to specified density. Thickness and type of pavement shall be as established by contract documents.

6.0 INSPECTION

6.1 GENERAL

The duty of the inspector is to examine the materials furnished and the work performed to verify full compliance with the contract documents. A dedicated, qualified inspector should be on-site to observe all phases of the site preparation, materials receiving and handling, installation, and site restoration. Observations of materials, workmanship, and, where specified in the contract documents, methods and means of performing construction are required to determine compliance with contract documents.

The inspector should have unrestricted access to all areas where the preparation of the materials and parts of work to be done are carried out and conducted. The contractor shall provide access to all facilities and assistance required to perform the inspection.

6.2 INSPECTION OF MATERIALS

All construction materials must be carefully and thoroughly inspected prior to and during placement. Inspection should be an ongoing process, since satisfactory materials first arriving on-site can be damaged during handling, storage, and installation. No material of any kind should be used prior to inspection and formal approval.

Project specifications and other product-specified information should form the basis for determining the suitability of all materials. Should any doubt concerning suitability arise, the product manufacturer should be consulted.

Shipments of select fill materials and drainage products should be accompanied by certified test reports. If such data are missing, laboratory tests should be used to confirm appropriate properties. All drainage products or drainage system components should be measured to check size, shape, and fit.

All materials must be inspected to ensure that they are free of foreign deposits, defects, and damage. Cleaning and removal of foreign matter may be acceptable, provided there is complete assurance that the construction material or product is unharmed and in "like new" condition.

Damaged products or components should be immediately removed from the site. Repairs may be performed on damaged goods following inspection and approval of the inspector and after consultation with the manufacturer. The return of any previously rejected materials, products, or components to the site is acceptable only after reinspection and approval following rework. Any material or workmanship found at any time during the construction cycle not in accordance with project specifications, for any reason, shall immediately be remedied.

6.2.1 Prefabricated and/or Premanufactured Components

Prior to installation, all prefabricated and/or premanufactured components shall be inspected to establish conformity with the project specifications and to check for damage and the presence of foreign matter. The manufacturer's certificate of compliance and product drawings should confirm compliance with the contract documents.

6.2.2 Bedding, Backfill, and Envelope Materials

All materials for use in bedding, backfilling, and envelope materials, or as otherwise used in subsurface drainage systems, shall be checked for conformance to project specifications. The supplier's certificate of compliance may form the basis of the inspection.

Geosynthetic products should be inspected for damage and conformity to the project specifications and drawings. The manufacturer's materials certificate should be the basic inspection document. Any deviations from the contract specifications should be immediately referred to the project engineer.

6.2.3 Storage of Materials

Storage of materials should be managed by the contractor to avoid impairing the usability and quality of on-site materials. Observance of any special handling methods required shall be verified and recorded. Storage or special protection required by the contract documents for certain items should also be verified. The inspector is responsible for monitoring the contractor's observance of these requirements.

6.3 INSPECTION OF EQUIPMENT

The contractor should verify that the equipment provided is appropriate in size for performance of the contract, is in good operating condition, and meets safety requirements.

6.4 INSPECTION OF CONSTRUCTION

6.4.1 General

The sequence of construction operations is an important consideration in projects that require construction in a particular order. Sequencing may also be required in order to allow the existing facilities to remain in operation during the construction phase. The contract documents normally allow as much flexibility as possible in sequencing operations and may have no requirements other than an overall completion date.

It is not the function of the inspector to supervise or direct the manner in which the work is performed. The inspector should follow each stage of construction so that any construction errors can be resolved during construction rather than after. The inspector should closely monitor all acceptance testing for the correct test procedure. Inspection activities associated with each construction stage are summarized in Table 6-1.

6.4.2 Construction Layout

Survey controls should be established as referenced on the contract drawings. It is the contractor's responsibility to stake and build the project from the controls. All necessary auxiliary staking must be in place prior to construction. If an error in auxiliary staking is observed, detected, or suspected, the error shall be promptly called to the contractor's attention.

It is the responsibility of the inspector to ascertain that the survey control points are in place as referenced. If there is any evidence that the control points have been disturbed, the inspector shall notify the engineer, who will arrange to have the points checked by the surveyor who originally set the points or take other appropriate action.

6.4.3 Excavation and Dewatering

The inspector should confirm that all excavations and dewatering activities are performed in accordance with the contract documents and that these activities will allow construction to be completed according to plan. Proper disposal of water is the responsibility of the contractor.

Trenches shall be excavated to depths and widths as specified for correct backfill and/or envelope placement and compaction. Water standing or flowing into the trench should be removed until backfill and envelope materials are placed. When possible, trenching should occur across slope and not downslope to minimize soil loss during rain.

Prior to placement of drainage products and envelope materials, all finished excavations should be inspected to ensure the absence of unsuitable materials.

6.4.4 Installation

Where the contract documents reference installation in accordance with the manufacturer's directions, such directions shall be provided for use in verifying that subsequent installation is performed in accordance with them.

6.4.5 Backfilling

The drain should be inspected for proper elevation, grade, alignment, and joint spacing; collapsed, broken, or cracked pipe; and thickness of aggregate envelope before backfilling. Backfill placement should be in accordance with Subsection 5.10.

6.4.6 Televising

The underdrain should be inspected by means of closed-circuit television or other acceptable camera systems where appropriate. Permanent videotape or film should be furnished in accordance with the contract documents.

6.4.7 Testing

There are few definitive tests that can be performed on installed subsurface drains that give measurable indications of the functional effectiveness of the installation.

The specifications may require specific field tests to be performed. As appropriate, samples shall be furnished by the contractor or representative samples will be taken from delivered materials. The number of samples shall be sufficient to satisfy all testing requirements. Control testing shall be performed in the field or at such other locations as required. The overall inspector, together with the engineer, will review the testing requirements and determine the overall testing program including all necessary testing facilities and record forms will be specified.

Complete records of the test and results shall be retained. Specimens shall be retained if they are important to prove results of the specified tests. Before being accepted as completed, each drain should be tested for obstructions and any deflections or deformations.

6.4.8 Safety

All safety rules established by governmental agencies shall be strictly followed. Observations should be performed in a manner that will not unreasonably impede or obstruct the contractor's operations.

6.4.9 Suitability and/or Conformance

Methods and means of construction are left to the option of the contractor on most items to allow flexibility, but they may be specified on items where methods and means are critical to obtaining a final desired product. While means and methods may be specified in full detail, minimum elements are frequently identified. Where methods and means are specified in the contract documents, verification is required for compliance.

6.5 ACCEPTANCE OF CONSTRUCTION

Acceptance of construction normally covers the entire job and is not done on an incremental basis. In normal day-to-day operations, field personnel may verbally acknowledge apparent compliance with the contract documents. However, such acknowledgment should not constitute acceptance of part or all of the construction.

6.6 RECORDING OBSERVATIONS

Work performed by the contractor on a shift basis should be recorded by the inspector to provide a detailed record of the progress. All observations of noncompliance with the contract documents shall be recorded in the daily report. The report should cover any verbal statements made to and by the contractor concerning the noncompliance. Photographs should be taken when they assist in describing the noncompliance.

On matters not immediately corrected, the inspector should give the contractor a separate written Notice of Noncompliance within 24 hours. The notice should state specifically why the work does not meet the requirements of the contract documents.

6.7 RECORD DRAWINGS

Record details of construction should be incorporated into a final revision of the construction drawings to represent the most reliable record for future use.

During construction, the contractor and/or inspector should measure, reference, and record the locations of all inlets, outlets, and stubs for future connections, and other buried facilities. All construction changes from the original plans should be recorded. Contract drawings should be revised to indicate the field information after the project is completed. A notation such as "Revised According to Field Construction Records" or "Record Drawing" should be made on each sheet along with the inspector's name, date, and company name. Records of such plans should become a part of the owner's permanent records.

For subsurface drainage systems, the following minimum information should be included on the record drawing:

- Size and type of all drains on plan and profile sheets.
- Station and pipe invert elevation of all tees, wyes, cleanouts, manholes, and outfalls.
- Manholes and other critical points referenced to established survey points.

Table 6-1.
Inspection Activities Associated with Construction

Construction Activity	Inspection Activity
1. Site preparation	Monitor clearing and grubbing and disposition of topsoil
2. Subgrade/foundation preparation	Observation and verification of grades
	Verification of suitability of subgrade
3. Receiving/storing materials	Verification of type and condition of materials received at site and storage procedures
4. Trench excavation	Observation and verification of width, locations, lines, and grades
5. Installation of subsurface drainage systems	Verification of lines and grades, and testing where necessary
6. Initial backfilling	Verification of proper filter layer placement
	Verification of use of proper placement and compaction technique
	Verification of specific gradation, thickness, and densities
7. Final backfilling	Verification of proper backfill material placement procedure and compaction procedure
	Test verification of specified density and moisture content
8. Site restoration	Verification of final site conditions as specified

Appendix A
Recommended Use of Trench Boxes

1.0 INTRODUCTION

Trench boxes provide a safer work area to install pipe in deep trenches or in soils that have insufficient stability. Use of a trench box may also be required by the specifications for reasons other than safety. While these guidelines will work for most cohesive and non-cohesive native soils, highly unusual soil conditions may require further investigation.

2.0 GENERAL CONSIDERATIONS

Some installations may not require trench boxes if the trench sidewall can be sloped. The engineer should provide specific guidance on acceptable slopes, but in no case should the trench wall slope be greater than the angle of repose of the native soil.

The length of the trench box should be suitable for the pipe length.

3.0 SUBTRENCH CONSTRUCTION

The most effective way to maintain a sound subtrench system is to provide a subtrench within which to place the pipe and backfill. Backfill and compact according to the design specifications within the subtrench. The trench box can be pulled along the top edge of the subtrench (Figure 3-1) without affecting the pipe or the backfill.

Subtrench construction also makes it easier to use a geotextile around the backfill if it is required by the project specifications. Line the subtrench with the geotextile, place the pipe and backfill, and wrap the geotextile over the top of the pipe and backfill system.

4.0 REGULAR TRENCH CONSTRUCTION

In construction not involving a subtrench situation, dragging a trench box subjects the pipe to stretching or separated joint. The box should be lifted vertically until it is above the pipe, and reset into its new position. If it is necessary for a trench box to be dragged through a trench, do not lower the box more than one-fourth of the nominal diameter below the crown (top) of the pipe. This allows the backfill material to flow out of the bottom of the box around the pipe so that backfill disturbance is kept to a minimum.

An alternative for flexible stormwater system when the box will be dragged is to use a well-graded granular backfill material two diameters on either side of the pipe and compact it to a minimum of 90% standard Proctor density before moving the box. Immediately fill the area between the pipe and backfill structure and the trench wall with a granular material.

If the project requires a geotextile around the backfill, use a well-graded granular backfill material and compact it to at least 90% standard Proctor density. Do not drag the box; instead, lift it vertically. After the trench box is removed, immediately fill

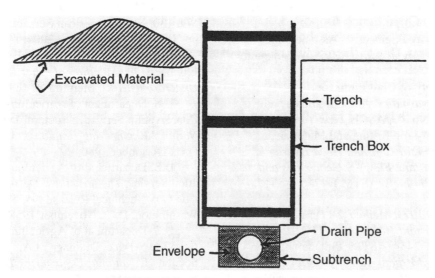

FIGURE 3-1. Subtrench Construction.

the area between the pipe and backfill structure and the trench wall with a granular material and compact according to project requirements. The geotextile manufacturer may be able to provide additional information regarding the suitability of specific geotextiles for use with trench boxes.

5.0 SUMMARY

While trench boxes increase worker safety in difficult site conditions, precautions are required to ensure a structurally sound pipe and backfill system.

Construction of a subtrench is the most effective means of maintaining a sound system. When a regular trench is used, techniques such as lifting the box, keeping the box about three-fourths the nominal diameter up from the trench bottom, and providing a wide granular backfill envelope will help provide a quality construction.

7.0 REFERENCES

7.1 REFERENCE DOCUMENTS

American Association of State Highway and Transportation Officials, *Standard Specifications for Highway Bridges*, Washington, D.C., 1992.

American Concrete Pipe Association, *Concrete Pipe Design Manual*, ACPA, Vienna, Va., 1992.

ACPA, *Concrete Pipe Handbook*, ACPA, Vienna, Va., 1988.

American Iron and Steel Institute, *Handbook of Steel Drainage and Highway Construction Products*, AISI, New York, 1971.

American Railway Engineering Association, *Manual for Railway Engineering*, AREA, Washington, D.C., 1993.

American Society of Civil Engineers, *Nomenclature for Hydraulics,* ASCE Manuals and Reports of Practice No. 43, ASCE, New York, 1962.

American Society for Testing and Materials, *Standard Test Method for Constant Head Hydraulic Transmissitivity (In-Plane Flow) of Geotextiles and Geotextile Related Products*, ASTM D4716, ASTM, Philadelphia, 1992.

ASTM, *Standard Test Method for Permeability of Granular Soils (Constant Head)*, AASHTO T215, E1-1993 R, ASTM D2434, ASTM, Philadelphia, 1994.

ASTM, *Standard Practice for Underground Installation of Thermoplastic Pipe for Sewers and Other Gravity-Flow Applications*, ASTM D2321-00, ASTM, Philadelphia, 2000.

Chambers, R. E., McGrath, T. J., and Heger, F. J., *Plastic Pipe for Subsurface Drainage of Transportation Facilities*, Transportation Research Board, National Cooperative Highway Research Program Report 225, Washington, D.C., October1980.

Federal Aviation Administration, U.S. Department of Transportation, *Airport Drainage*, ACI 50/5230-5B, Washington, D.C., 1970.

Federal Highway Administration, *Durable Pavement Systems, Participants' Notebook*, Office of Technology Applications and Office of Engineering, Washington, D.C., 1992.

Graber, S. D. Collection Conduits Including Subsurface Drains. *Journal of Environmental Engineering*, Vol. 130, No. 1, American Society of Civil Engineers, pp. 67–80, January 2004.

NRCS, SCS National Engineering, *Field Handbook*, Subchapter C. Sect. 650, 1428 (b), USDA, NRCS, Washington, D.C., 20250-0016, 1991.

National Clay Pipe Institute, *Clay Pipe Engineering Manual*, NCPI, Washington, D.C., 1982.

Sherard, J. L., Dunnigan, L. P., and Talbot., J. R., Basic Properties of Sand and Gravel Filters*, Journal of Geotechnical Engineering*, Vol. 110, No. 6, American Society of Civil Engineers, pp. 684–700, June 1984a.

Sherard, J. L., Dunnigan, L. P., and Talbot, J. R., Filters for Silts and Clays. *Journal of Geotechnical Engineering*, Vol. 110, No. 6, American Society of Civil Engineers, pp. 701–718, June 1984b.

Spangler, M. G., *Soil Engineering,* International Textbook Co., Scranton, Penn., 1966.

TeKrony, R. G., Sanders, G. D., and Cummins, B., History of Drainage in the Bureau of Reclamation, *Journal of Irrigation & Drainage Engineering* , Vol. 130, No. 2, American Society of Civil Engineers, pp. 148–153, March/April 2004.

U.S. Army Eng. Waterways Exp. Sta., Corps of Engineers, *Investigation of Filter Requirements for Underdrains*, Tech Memo. No. 183-1, 35 p., 1941.

U.S. Department of Agriculture, Soil Conservation Service, Structural Design, Section 6, *National Engineering Handbook,* USDA, Washington, D.C., December 1980.

U.S. Department of Agriculture, Soil Conservation Service, *The Structural Design of Underground Conduits,* Technical Release No. 5, USDA, Washington, D.C., November 1958.

U.S. Department of Agriculture, Soil Conservation Service, Drainage of Agricultural Land, Section 16, *National Engineering Handbook*, USDA, Washington, D.C., 1971.

U.S. Department of the Interior, Bureau of Reclamation, *Drainage Manual*, USDI, Denver, Colo., 1993.

U.S. Department of the Interior, Bureau of Reclamation, *Water Measurement Manual*, USDI, 1997.

Uni-Bell PVC Pipe Association, *Handbook of PVC Pipe: Design and Construction*, Dallas, 1986.

Wyatt, T., Baker, W., and Hall, J., *Drainage Requirements in Pavements—DRIP*, FHWA-SA-96-070, January 1998.

7.2 GENERAL REFERENCES

Advanced Drainage Systems, Inc., *Specifier Manual*, Columbus, Ohio, 1984.

American Concrete Pipe Association, *Concrete Pipe Installation Manual*, ACPA, Vienna, Va., 1988.

American Society of Agricultural Engineers, *Hydrologic Modeling of Small Watersheds*, ASAE, St. Joseph, Mich., 1982.

American Society for Testing and Materials, *Standard Test Method for Compressive Properties of Rigid Cellular Plastics*, ASTM D1621, ASTM, Philadelphia, 1992.

ASTM, *Standard Practice for Underground Installation of Thermoplastic Pipe for Sewers and Other Gravity-Flow Applications,* ASTM D2321-00, ASTM, 2000.

American Society of Civil Engineers, *Gravity Sanitary Sewer Design and Construction*, ASCE Manuals and Reports on Engineering Practice No. 60, ASCE, New York, 1982.

ASCE, "In-Plane Composite Drains," *Civil Engineering*, ASCE, New York, pp. 48–51, August 1984.

ASCE, *Urban Subsurface Drainage*, ASCE Manuals and Reports on Engineering Practice No. 95, ASCE, New York, 1998.

Anderson, B., *Underground Waterproofing*, WEBCO, Stillwater, Minn., 1983.

Beck, D. E., Testing and Comparing Geocomposite Drainage Products, *Geotechnical Fabrics Report*, Industrial Fabrics Association International, St. Paul, Minn., July/August 1988.

Bouwer, H., *Groundwater Hydrology*, McGraw-Hill, New York, 1978.

Cedergren, H. R., *Seepage, Drainage and Flow Nets*, John Wiley & Sons, New York, 1967.

Cedergren, H. R., *Drainage of Highways and Airfield Pavements,* John Wiley & Sons, New York, 1974.

Chamber's Technical Dictionary, 3rd Edition, Macmillan Co., New York, 1967.

Davis, C. V. and Sorensen, K. E., *Handbook of Applied Hydraulics*, McGraw-Hill, New York, 1986.

Davis, S. N. and DeWiest, R. J. M., *Hydrogeology*, John Wiley & Sons, New York, 1966.

Dempsey, B. J., *Pavement Drainage System Design*, prepared for Wisconsin DOT, February 15–16, 1988.

Driscoll, F. G., Ed., *Ground Water and Wells,* 2nd Edition, Johnson Division UOP, St. Paul, Minn., 1986.

Engineers Joint Council, *Thesaurus of Engineering and Scientific Terms*, New York, December 1967.

Federal Highway Administration, *Evaluation of Test Methods and Use Criteria for Geotechnical Fabrics in Highway Applications*, Report No. FHWA/RD-80-021, Washington, D.C., 1980.

FHA, *Design of Urban Highway Drainage*, Report No. FHWA-TS-79-225, Washington, D.C., 1983.

FHA, *Highway Subdrainage Design*, Report No. FHWA-TS-224, Washington, D.C., 1980.

FHA, *Hydraulic Design of Highway Culverts*, Hydraulic Design Series No. 5, Washington, D.C., 1985.

FHA, *Design of Highway Drainage—The State of the Art,* Report No. FHWA-TS-79-225, Washington, D.C., 1979.

Fetter, C. W., Jr., *Applied Hydrogeology*, Charles E. Merrill, Columbus, Ohio, 1980.

Freeze, R. A. and Cherry, J. A., *Groundwater*, Prentice-Hall, Englewood Cliffs, N.J., 1979.

Geosystems, Inc., Vertical Drains, *GeoNotes—A Ground Improvements Update,* Sterling, Va., undated.

Hancor, Inc., *Recommended Installation Practice for Hancor Hi-Q Titelines, Heavy Duty and Heavy Duty-AASHTO Pipe*, Findlay, Ohio, 1993.

Hannon, J. D. and California DOT, *Underground Disposal of Storm Water Runoff, Design Guidelines Manual*, FHWA-TS-80-218, Federal Highway Administration, Washington, D.C., 1980.

Hem, J. D., *Study and Interpretation of the Chemical Characteristics of Natural Water,* U.S. Geological Survey Water-Supply Paper 1473, Washington, D.C., 1970.

Illinois Department of Transportation, *Highway Standards Manual*, Springfield, Ill., November 1993.

Koerner, R. M., *Designing with Geosynthetics,* 2nd Edition, Prentice-Hall, Englewood Cliffs, N.J., 1990.

Lafayette Farm & Industry, *Agri-Fabric Awareness Manual,* Cuba City, Wis., undated.

Lohman, S. W. et al., *Definitions of Selected Ground Water Terms-Revisions and Conceptual Refinements,* Geological Survey Water-Supply Paper 1988, U.S. Geological Survey, Washington, D.C., 1972.

Merritt, F. S., *Standard Handbook for Civil Engineers,* McGraw-Hill, New York, 1983.

Peck, R. B., Hanson, W. E., and Thornburn, T. H., *Foundation Engineering,* John Wiley & Sons, New York, 1974.

Powers, J. P., *Construction Dewatering,* John Wiley & Sons, New York, 1979.

Roister, D. L., *Landslide Remedial Measures,* Tennessee Department of Transportation, Nashville, Tenn., 1982.

Sacks, A. M., "Geosynthetics," p. 14, *Remodeling Magazine,* Hanley Woods Inc., Washington, D.C., November 1987.

Sacks, A., "R_x for Basement Water Problems," *The Family Handyman,* St. Paul, Minn., pp. 36–40, September 1981.

Schuster, R. L. and Kruse, R. J., eds. *Landslides Analysis and Control, Transportation Research Board,* Special Report 176, Washington, D.C., 1978.

Schwab, G. O., Revert, R. K., et al., *Soil and Water Conservation Engineering,* 3rd Edition, John Wiley & Sons, New York, 1981.

Sowers, G. F., *Introductory Soil Mechanics and Foundations: Geotechnical Engineering,* Macmillan Publishing Co., New York, 1979.

Todd, D. K., *Ground Water Hydrology,* John Wiley & Sons, New York, 1980.

U.S. Department of Agriculture, "Soil Conservation Service, Drainage," Chapter 14 in: *Engineering Field Manual,* USDA, Washington, D.C., 1969.

USDA, *Soil Conservation Service, Technical Guide,* Section IV, Standard 606, Subsurface Drainage, May 1988.

U.S. Department of the Interior, Bureau of Reclamation, Ground Water Manual, Washington, D.C., 1995.

Vetch, J. O. and Humphrey, C. Rs, *Water and Water Use Terminology,* Thomas Printing & Publishing Co., New York, 1966.

Standard Guidelines for the Operation and Maintenance of Urban Subsurface Drainage

Contents

FOREWORD

The *Standard Guidelines for the Operation and Maintenance of Urban Subsurface Drainage* is an independent document intended to complement the ASCE Manuals and Reports on Engineering Practice No. 95, *Urban Subsurface Drainage*. These standard guidelines are companions to the *Standard Guidelines for the Design of Urban Subsurface Drainage* and *Standard Guidelines for the Installation of Urban Subsurface Drainage*. These standard guidelines were developed by the Urban Drainage Standards Committee, which is responsible to the Environmental and Water Resources Institute of the American Society of Civil Engineers.

The material presented in this publication has been prepared in accordance with recognized engineering principles. These standard guidelines should be used only under the direction of an engineer who is competent in the field of urban subsurface drainage. The publication of the material contained herein is not intended as a representation or warranty on the part of the American Society of Civil Engineers, or of any other person named herein, that this information is suitable for any general or particular use, or promises freedom from infringement of any patent or patents. Anyone making use of this information assumes all liability from such use.

ACKNOWLEDGEMENTS

The American Society of Civil Engineers (ASCE) acknowledges the work of the Urban Drainage Standards Committee of the Environmental and Water Resources Institute of ASCE (EWRI of ASCE).

William Curtis Archdeacon, Chair
Richard H. Berich
Christopher B. Burke
Robert T. Chuck
F. Scott Dull
Robert S. Giurato, Secretary
S. David Graber
Jay M. Herskowitz
Conrad G. Keyes, Jr.
John M. Kurdziel
John J. Meyer
Philip M. Meyer
James R. Noll

This group comprises individuals from many backgrounds, including consulting engineering, research, the construction industry, education, and government. Those individuals who serve on the Urban Drainage Standards Committee are:

Walter J. Ochs
Garvin J. Pederson
Glen D. Sanders
Erez Sela
Alan N. Sirkin
Edward L. Tharp
William J. Weaver
Richard D. Wenberg, Past Chair
David L. Westerling
Stan E. Wildesen
Lyman S. Willardson (deceased)
Donald E. Woodward

Standard Guidelines for the Operation and Maintenance of Urban Subsurface Drainage

1.0 SCOPE

The intent of this standard is to present operation and maintenance guidance for urban subsurface drainage systems. The collection and conveyance of subsurface drainage waters are within the purview of this standard for applications such as airports, roads, and other commercial transportation systems, and industrial, residential, and recreational areas. Incidental surface water is considered. This document is intended for guidance to the owner during the operational phase.

This standard does not address agricultural drainage, landfills, recharge systems, detention ponds, conventional storm sewer design, or the use of injection systems.

Customary units and standard international (SI) units are used throughout this document.

2.0 DEFINITIONS

2.1 GENERAL

This section defines specific terms for use in these guidelines. The reference documents listed in Section 3.0 may also be helpful in understanding terms in this standard.

2.2 TERMS

Aquifer—Water-bearing stratum of permeable rock, sand, or gravel.

Bedding—Support for pipe during the construction process.

Drain envelope—Generic name for materials placed on or around a drainage product, irrespective of whether used for mechanical support (bedding), hydraulic purposes (hydraulic envelope), or to stabilize surrounding soil material (filter envelope).

EPA—Environmental Protection Agency.

Filter envelope—Permeable material placed around a drainage product to stabilize the structure of the surrounding soil material. A filter envelope may initially allow some fines and colloidal material to pass through it and into the drain.

Geocomposite—Geosynthetic materials for collecting and transporting water while maintaining soil stability.

Geomembrane—Sheet materials intended to form an impervious barrier.

Geosynthetic—Synthetic material or structure used as an integral part of a project, structure, or system. Within this category are subsurface drainage and water control materials such as geomembranes, geotextiles, and geocomposites.

Geotextile—Woven or nonwoven engineering fabric sheet material intended to allow the passage of water (but not fines, and without collecting fines at the soil-textile interface).

Hydraulic envelope—Permeable material placed around a drainage product to improve flow conditions in the area immediately adjacent to the drain.

Inspection—Critical examination to determine conformance to applicable quality standards or specifications.

Iron ochre—Red or yellow gelatinous deposit formed by a combination of soluble iron deposits and bacteria.

Jetting—Method used to clean pipes involving high-pressure water.

OSHA—Occupational Safety and Health Administration, the federal agency responsible for safety and health concerns on construction job sites.

Rodding—Method used to clean pipes involving mechanical means.

Record Drawing—Drawings prepared during or after construction showing the final measurements of construction, including any deviations from the design drawings and certain other field observations such as tie-in locations.

Subsurface water—All water beneath the ground or pavement surface. Sometimes referred to as groundwater.

3.0 OPERATION AND MAINTENANCE PLAN

3.1 GENERAL

This section is a generalized outline of what should be included in an operation and maintenance plan. It may be necessary to modify this plan to more accurately reflect the specific subsurface drainage system under consideration.

This is intended as a general standard guideline for preparing procedures and timetables related to routine operation and maintenance of subsurface drainage systems. Technical personnel must be familiar with the basic concepts of subsurface drainage facilities.

Operation and maintenance instruction materials submitted by manufacturers supplying equipment for urban subsurface system components should be retained and incorporated into the procedures document. The manual should explain the general operational relationships between the various system components of the facility and include any manufacturers' instructions or recommendations. Adherence to these procedures is essential to retain the operating capacity of a facility throughout its expected service life.

3.2 SCOPE

An operations and maintenance plan is intended to provide general guidance for the topics listed. However, this list of subjects may not cover all aspects of any particular subsurface drainage system.

- Normal operating procedures to be followed during the majority of the facility's service life.
- Special operating procedures to be followed when normal procedures cannot be followed.
- Routine maintenance procedures specifically described according to a definite schedule.
- Safety procedures for all operating personnel.
- Emergency procedures for conditions involving serious service disruptions.

3.3 RESPONSIBILITIES

The owner has the responsibility for the operation and maintenance of the subsurface drainage system. Generally, these areas of responsibility include:

- Providing adequate funds for supplies and personnel, operation and maintenance of equipment, and any necessary system expansion.
- Selecting either trained personnel or providing training and education for qualified individuals.
- Developing and implementing a comprehensive program that includes a complete reporting and records retention system.
- Coordinating with other departments for maintenance and operation to ensure an integrated system is working.

3.4 DESIGN CRITERIA

The procedures manual should detail the performance criteria used in the design. All personnel should be aware of these criteria so that deviations can be recognized.

3.5 NORMAL OPERATING PROCEDURES

Normal operation is the design condition expected of the system. When all components are functioning as designed, the only requirements of the operating personnel are routine checks and scheduled maintenance.

3.6 ABNORMAL OPERATING PROCEDURES

3.6.1 General
There are several types of abnormal operating procedures to consider.

3.6.2 Line Blockage
Subsurface drains are often designed for minimal flow. As a result, subsurface drains are susceptible to blockage. If a pipeline becomes blocked, rodding or jetting may be required to clear it. During cleaning operations, a careful watch should be maintained at the downstream manhole for an indication of the cause of the blockage.

Cleanout assemblies may be located at periodic intervals along the pipeline as shown on the "as-built/record drawings." If a conveyance main becomes blocked, the nearest of these cleanouts can be located, accessed, and cleaning equipment inserted.

Hydraulic flushing may be attempted to clean any pipeline. Special care should be taken to avoid damage caused by hydraulic surging.

3.6.3 Other System Components
Other components should be operated according to manufacturers' recommendations. In the event of reduced performance of system components, correction procedures in compliance with manufacturers' recommendations should be used.

3.7 MAINTENANCE PROCEDURES

3.7.1 General
Maintenance can be broadly classified as either corrective or preventive.

3.7.2 Corrective Maintenance
Corrective maintenance involves the repair of equipment after breakdown. The operation and maintenance plan should include all necessary equipment manuals, diagrams, and instructions for satisfactory operation.

3.7.3 Preventive Maintenance

Equipment breakdown is often related to a failure of preventive maintenance. As the term implies, preventive maintenance is intended to prevent disruptive breakdowns. Since many components that require different preventive maintenance actions at different time intervals may be involved in the system, preventive maintenance is best performed on a scheduled basis from a checklist. Manufacturers' recommendations should be followed during preventive maintenance.

3.7.4 Checklists

Checklists should be maintained for each component of the urban subsurface system with recommended schedules for inspection clearly stated. Checklists should include each aspect of the inspection such as: damage to structure, evidence of restricted capacity, and manufacturers' recommended maintenance. Once all inspections outlined on the form have been completed, the forms should be replaced and the completed forms filed in the owner's records.

3.7.5 Annual System Inspection

A general overview inspection of the entire system should be performed at least annually in addition to the more intensive inspections performed as discussed in Section 6.0 of these standard guidelines. Local conditions may make more frequent inspections of certain components necessary. These may be adjusted as experience is gained in the operation of the system, but in no case should the annual inspection be extended.

3.8 SAFETY

3.8.1 General

All personnel are responsible for keeping areas safe and clean. Guards should be in place on operating equipment, and all areas should be properly lighted. All enclosed space should be adequately ventilated prior to personnel entering. All personnel should be sure they understand the following:

- Location of all safety equipment.
- Use of safety equipment and devices.
- OSHA, EPA, and local safety rules.
- Potential for "danger."

3.8.2 Mechanical

When working on pumps, suction and discharge valves are fully closed. Maintenance on equipment in operation should be limited to lubrication, packing adjustments, minor repair, or as allowed by the manufacturer's instructions.

3.8.3 Electrical

Lock out and tag main switches of electrical equipment before beginning work. Do not remove any tag without first checking with the person who initiated the tag. Any unusual motor temperature, noise, vibration, and so on should be reported and logged.

3.8.4 Underground Procedures

Where excavation of underground facilities is undertaken, OSHA-approved trenching procedures must be followed. Sloping or shoring requirements vary with local geologic and soil conditions. Requirements for sloping, benching, and shoring are found in OSHA 29CFR1926.652, appendices A, B, C, and D. When excavating, the operators must be alert to both overhead power lines and underground utilities that pose a hazard to excavation equipment or personnel. Buried gas and electric lines pose the greatest hazard to operators, but accidents involving communication lines, water lines, and other utilities can be very costly to those causing the damage. Telephone numbers for all underground utility locators should be included in the operation and maintenance plan.

3.8.5 Other Safety Considerations

Workers entering structures must realize that surface drainage systems are likely to be part of the structure and subject to high flow conditions. Workers entering during these times should be protected by safety lines, flotation devices, and lighting. System features and signage designed to provide public safety should be kept in good condition. For further information, refer to OSHA safety requirements for workers in confined spaces.

4.0 WATER QUALITY

4.1 GENERAL

Periodic observation of the flow is necessary to monitor possible water quality degradation. The potential for pollutants to be present is constantly changing. Improvements and/or developments within the drainage basin can generate substances having a pollution potential that could be conveyed to the subsurface drainage system.

An increase in flow rate is often accompanied by higher velocity, which is more effective in transporting pollutants. This, in fact, could be the only cause for concern of such pollutants as suspended solids and/or turbidity. Quantity and quality should be monitored simultaneously.

EPA and other local regulations affect the allowable water quality of subsurface drainage. The water quality requirements will affect the design of the system.

4.2 ENVIRONMENTAL INDICATORS

A review of the area should be performed to determine any changes since the construction of the subsurface system. These changes will then have to be evaluated as to possible effects on the subsurface flow. Water sampling of aquifers and watershed sources representing existing and potential sources of subsurface water supply may be required. Certain parameters and their background levels can be expected to occur naturally in the water due to the existing environment. By visual inspection or through personal observation, a determination can be made for the necessity and extent of a field sampling program. If test results show unusual concentrations or unexpected constituents in the water, further investigations could be necessary. A treatment program may need to be implemented, or modifications may need to be proposed that would mitigate or eliminate adverse impacts caused by the problem constituents.

4.3 WATER QUALITY STANDARDS

Limitations may be imposed on the discharge based on the receiving waters. Water quality standards limit the concentration of various parameters to be discharged. Standards are generally established by the federal, state, or local governments and are subject to continual revisions. These standards can be used as a basis for evaluating water quality. The suggested list of parameters for analysis include, as a minimum: temperature, color, odor, pH, dissolved oxygen, total suspended solids, and turbidity. Considerations include inorganic chemicals, heavy metals, corrosiveness, organic chemicals (pesticides/herbicides), and microbiological and radioactive materials.

Sampling points and frequency can best be determined in the field. Sampling points should be located at strategic points throughout the system.

5.0 PERIODIC INSPECTION

5.1 GENERAL

Most systems experience long-term flow capacity reductions. Maximum subsurface drainage system performance requires maintenance programs that keep systems clean, soil-tight, structurally intact, and free of debris.

Routine frequent inspections of all subsurface drainage systems are necessary to ensure the continuing level of water management intended by the initial planning and design. A uniform inspection schedule applicable to all sites cannot be established due to regional variations in climate or environment and local conditions affecting land usage. Experiences with similar systems may indicate specific local problem areas that should become a formal part of an inspection program.

Record keeping is important. Complete records of previous inspections provide a gauge for comparison to determine the rate and severity of deterioration.

5.2 UNDERGROUND SAFETY

All persons inspecting conduits below the ground surface or any other confined space in the system should take the utmost caution when entering these areas. Deadly gases such as methane can be produced by decaying material within the storm drain or these gases can seep in from adjacent sewers or gas lines. All subsurface drains should be considered suspect even though they have numerous air inlets to the surface. All persons entering underground conduits or other confined spaces are required to conform to OSHA regulations as published in the Federal Register, Volume 58, Number 9/Thursday, January 14, 1993/Rules and Regulations for entry into confined spaces.

5.3 INSPECTION

Inspections should determine the current operating and structural status of the subsurface drainage system and provide thorough examination for symptoms of developing problems that can alter future performance. Typical elements comprising a thorough inspection are:

1. Check for accumulations of debris, rodents, or other flow impediments at inlets and outlets. Flap gates and other water control devices should operate freely.
2. Inspect the system interior, if possible, for tree or other vegetation roots, mineral deposits, trash or silt accumulations, and other foreign objects obstructing flow paths. Often, televising the system is required to provide this service.

3. Evaluate ground surfaces for evidence of subsurface drainage system leakage. Excessive groundwater from inoperable subsurface drainage systems may cause prematurely distressed pavements, loss of healthy vegetation, and topsoil saturations.
4. Examine inlet and outlet areas for evidence of soil erosion, which generally leads to scour, undermining, and caving of adjacent soils supporting the subsurface drainage system. Soil erosion quickly leads to reduced structural and hydraulic performance.
5. Inspect all visible structures such as catch basins, headwalls, culverts, and outlet pipes for signs of wear or breakage.
6. Check upstream for evidence of backups or prolonged surface water presence that indicates reduced inflow. Check downstream for evidence of foreign materials that indicate reduced filtration of soils, diminished screening of foreign particles, or structural degradation of the drainage system itself.

Visual inspection methods are frequently enhanced by electronic and optical aids such as television cameras, fiber-optic scopes, and laser beam equipment. These inspection tools reveal cracks, displacements and misalignments, and other interior problems with minimal disturbance to the subsurface drainage system.

6.0 MAINTENANCE

6.1 GENERAL

Maintenance of subsurface drainage facilities is often neglected. Most of the facilities are out of sight, so there is little public pressure to carry on maintenance programs. Failures tend to develop gradually, as opposed to being catastrophic, so that a failure may not be recognized for several years after it has occurred. Costs of failure in terms of property damage and reconstruction can be enormous compared to the cost of routine preventive maintenance. Therefore, it is usually in the best interest of the owner to develop and implement an aggressive preventive maintenance program. Structures such as manholes or outlet pipes should be checked on a regular schedule for signs of structural distress and loss of hydraulic function. Appropriate repairs should be made on a timely basis.

6.2 CLEANING

Methods and schedules for cleaning will depend on the type of subsurface drain being cleaned. The function of the drain, materials from which it is made,

susceptibility to clogging, and type of clogging (roots, trash, chemical, or biological) will all need to be considered. High-pressure hydraulic drain cleaners are satisfactory for most applications where access to the drain lines is provided. Initially, annual cleaning should be sufficient for roots and chemical or biological accumulations. As experience with the drain system is gained, the cleaning schedule should be adjusted accordingly.

Some installations, such as sport field drain grids or highway shoulder drains, may not provide access for mechanical cleaning.

Chemical treatment may be required in place of, or in conjunction with, mechanical cleaning. Chemical cleaning must be done in an environmentally responsible manner. For instance, iron ochre deposits, which can clog the openings in the pipe or restrict the carrying capacity of the pipe, sometimes require strong acid solutions for removal. These solutions must be contained until they are neutralized to the approximate pH of the receiving body of water.

6.3 MECHANICAL AND ELECTRICAL

Pumps and motors will usually have manufacturer-recommended maintenance guides. When they do not, the design engineer should provide one, particularly if there are any unusual conditions. Appurtenances, such as covers, valves, or flap gates, may also require occasional attention. Flap gates that are removed for maintenance of the drain should always be immediately replaced. If frequent access is required, a different design of flap gate may be required to avoid losing the original one.

6.4 REPAIR

Repairs to subsurface drains should be made using original materials to the extent possible. Since subsurface drains are designed to last up to 100 years, original materials may not be available. When substitutions are made, care should be taken to ensure that the capacity of the drain is not diminished.

The same safety requirements should be in effect for repairs as for original construction. Maintenance crews must avoid the temptation to take shortcuts because of the short duration of their activities. Cave-ins and toxic fumes are even more prevalent in repair work than during original construction. Repairs to electrical equipment should be done under standard electrical safety rules.

6.5 REHABILITATION

Rehabilitation of a subsurface drain system may involve complete cleaning of all components; realigning portions of the system; replacing worn or outdated electrical components; replacing gates, valves, and other appurtenances; repairing or replacing major structures, such as manholes and outlets; replacing entire sections of drains; or additions to the original system. In general, rehabilitation should follow the same overall pattern as original construction, including engineering analyses, cost comparison, up-to-date materials selection, construction methods, and safety procedures.

7.0 REFERENCES

7.1 REFERENCE DOCUMENTS

American Association of State Highway and Transportation Officials, *Standard Specifications for Highway Bridges*, Washington, D.C., 1992.

American Concrete Pipe Association, *Concrete Pipe Design Manual*, ACPA, Vienna, Va., 1992.

ACPA, *Concrete Pipe Handbook*, ACPA, Vienna, Va., 1988.

American Iron and Steel Institute, *Handbook of Steel Drainage and Highway Construction Products*, AISI, New York, 1971.

American Railway Engineering Association, *Manual for Railway Engineering*, AREA, Washington, D.C., 1993.

American Society of Civil Engineers, *Nomenclature for Hydraulics*, ASCE Manuals and Reports of Practice No. 43, ASCE, New York, 1962.

American Society for Testing and Materials, *Standard Test Method for Constant Head Hydraulic Transmissitivity (In-Plane Flow) of Geotextiles and Geotextile Related Products*, ASTM D4716, ASTM, Philadelphia, 1992.

ASTM, *Standard Test Method for Permeability of Granular Soils (Constant Head)*, AASHTO T215, E1-1993 R, ASTM D2434, ASTM, Philadelphia, 1994.

ASTM, *Standard Practice for Underground Installation of Thermoplastic Pipe for Sewers and Other Gravity-Flow Applications*, ASTM D2321-00, ASTM, Philadelphia, 2000.

Chambers, R. E., McGrath, T. J., and Heger, F. J., *Plastic Pipe for Subsurface Drainage of Transportation Facilities*, Transportation Research Board, National Cooperative Highway Research Program Report 225, Washington, D.C., October 1980.

Federal Aviation Administration, U.S. Department of Transportation, *Airport Drainage*, ACI 50/5230-5B, Washington, D.C., 1970.

Federal Highway Administration, *Durable Pavement Systems, Participants' Notebook*, Office of Technology Applications and Office of Engineering, Washington, D.C., 1992.

Graber, S. D. Collection Conduits Including Subsurface Drains. *Journal of Environmental Engineering*, Vol. 130, No. 1, American Society of Civil Engineers, pp. 67–80, January 2004.

NRCS, SCS National Engineering, *Field Handbook*, Subchapter C. Sect. 650, 1428 (b), USDA, NRCS, Washington, D.C., 20250-0016, 1991.

National Clay Pipe Institute, *Clay Pipe Engineering Manual*, NCPI, Washington, D.C., 1982.

Sherard, J. L., Dunnigan, L. P., and Talbot., J. R., Basic Properties of Sand and Gravel Filters, *Journal of Geotechnical Engineering*, Vol. 110, No. 6, American Society of Civil Engineers, pp. 684–700, June 1984a.

Sherard, J. L., Dunnigan, L. P., and Talbot, J. R., Filters for Silts and Clays. *Journal of Geotechnical Engineering*, Vol. 110, No. 6, American Society of Civil Engineers, pp. 701–718, June 1984b.

Spangler, M. G., *Soil Engineering*, International Textbook Co., Scranton, Penn., 1966.

TeKrony, R. G., Sanders, G. D., and Cummins, B., History of Drainage in the Bureau of Reclamation, *Journal of Irrigation & Drainage Engineering*, Vol. 130, No. 2, American Society of Civil Engineers, pp. 148–153, March/April 2004.

U.S. Army Eng. Waterways Exp. Sta., Corps of Engineers, *Investigation of Filter Requirements for Underdrains*, Tech Memo. No. 183-1, 35 p., 1941.

U.S. Department of Agriculture, Soil Conservation Service, Structural Design, Section 6, *National Engineering Handbook*, USDA, Washington, D.C., December 1980.

U.S. Department of Agriculture, Soil Conservation Service, *The Structural Design of Underground Conduits*, Technical Release No. 5, USDA, Washington, D.C., November 1958.

U.S. Department of Agriculture, Soil Conservation Service, Drainage of Agricultural Land, Section 16, *National Engineering Handbook*, USDA, Washington, D.C., 1971.

U.S. Department of the Interior, Bureau of Reclamation, *Drainage Manual*, USDI, Denver, Colo., 1993.

U.S. Department of the Interior, Bureau of Reclamation, *Water Measurement Manual*, USDI, 1997.

Uni-Bell PVC Pipe Association, *Handbook of PVC Pipe: Design and Construction*, Dallas, 1986.

Wyatt, T., Baker, W., and Hall, J., *Drainage Requirements in Pavements—DRIP*, FHWA-SA-96-070, January 1998.

7.2 GENERAL REFERENCES

Advanced Drainage Systems, Inc., *Specifier Manual*, Columbus, Ohio, 1984.

American Concrete Pipe Association, *Concrete Pipe Installation Manual*, ACPA, Vienna, Va., 1988.

American Society of Agricultural Engineers, *Hydrologic Modeling of Small Watersheds*, ASAE, St. Joseph, Mich., 1982.

American Society for Testing and Materials, *Standard Test Method for Compressive Properties of Rigid Cellular Plastics*, ASTM D1621, ASTM, Philadelphia, 1992.

ASTM, *Standard Practice for Underground Installation of Thermoplastic Pipe for Sewers and Other Gravity-Flow Applications*, ASTM D2321-00, ASTM, 2000.

American Society of Civil Engineers, *Gravity Sanitary Sewer Design and Construction*, ASCE Manuals and Reports on Engineering Practice No. 60, ASCE, New York, 1982.

ASCE, "In-Plane Composite Drains," *Civil Engineering*, ASCE, New York, pp. 48–51, August 1984.

ASCE, *Urban Subsurface Drainage*, ASCE Manuals and Reports on Engineering Practice No. 95, ASCE, New York, 1998.

Anderson, B., *Underground Waterproofing*, WEBCO, Stillwater, Minn., 1983.

Beck, D. E., Testing and Comparing Geocomposite Drainage Products, *Geotechnical Fabrics Report*, Industrial Fabrics Association International, St. Paul, Minn., July/August 1988.

Bouwer, H., *Groundwater Hydrology*, McGraw-Hill, New York, 1978.

Cedergren, H. R., *Seepage, Drainage and Flow Nets*, John Wiley & Sons, New York, 1967.

Cedergren, H. R., *Drainage of Highways and Airfield Pavements*, John Wiley & Sons, New York, 1974.

Chamber's Technical Dictionary, 3rd Edition, Macmillan Co., New York, 1967.

Davis, C. V. and Sorensen, K. E., *Handbook of Applied Hydraulics*, McGraw-Hill, New York, 1986.

Davis, S. N. and DeWiest, R. J. M., *Hydrogeology*, John Wiley & Sons, New York, 1966.

Dempsey, B. J., *Pavement Drainage System Design*, prepared for Wisconsin DOT, February 15–16, 1988.

Driscoll, F. G., Ed., *Ground Water and Wells*, 2nd Edition, Johnson Division UOP, St. Paul, Minn., 1986.

Engineers Joint Council, *Thesaurus of Engineering and Scientific Terms*, New York, December 1967.

Federal Highway Administration, *Evaluation of Test Methods and Use Criteria for Geotechnical Fabrics in Highway Applications*, Report No. FHWA/RD-80-021, Washington, D.C., 1980.

FHA, *Design of Urban Highway Drainage*, Report No. FHWA-TS-79-225, Washington, D.C., 1983.

FHA, *Highway Subdrainage Design*, Report No. FHWA-TS-224, Washington, D.C., 1980.

FHA, *Hydraulic Design of Highway Culverts*, Hydraulic Design Series No. 5, Washington, D.C., 1985.

FHA, *Design of Highway Drainage—The State of the Art*, Report No. FHWA-TS-79-225, Washington, D.C., 1979.

Fetter, C. W., Jr., *Applied Hydrogeology*, Charles E. Merrill, Columbus, Ohio, 1980.

Freeze, R. A. and Cherry, J. A., *Groundwater*, Prentice-Hall, Englewood Cliffs, N.J., 1979.

Geosystems, Inc., Vertical Drains, *GeoNotes—A Ground Improvements Update*, Sterling, Va., undated.

Hancor, Inc., *Recommended Installation Practice for Hancor Hi-Q Titelines, Heavy Duty and Heavy Duty-AASHTO Pipe*, Findlay, Ohio, 1993.

Hannon, J. D. and California DOT, *Underground Disposal of Storm Water Runoff, Design Guidelines Manual*, FHWA-TS-80-218, Federal Highway Administration, Washington, D.C., 1980.

Hem, J. D., *Study and Interpretation of the Chemical Characteristics of Natural Water*, U.S. Geological Survey Water-Supply Paper 1473, Washington, D.C., 1970.

Illinois Department of Transportation, *Highway Standards Manual*, Springfield, Ill., November 1993.

Koerner, R. M., *Designing with Geosynthetics*, 2nd Edition, Prentice-Hall, Englewood Cliffs, N.J., 1990.

Lafayette Farm & Industry, *Agri-Fabric Awareness Manual*, Cuba City, Wis., undated.

Lohman, S. W. et al., *Definitions of Selected Ground Water Terms-Revisions and Conceptual Refinements*, Geological Survey Water-Supply Paper 1988, U.S. Geological Survey, Washington, D.C., 1972.

Merritt, F. S., *Standard Handbook for Civil Engineers*, McGraw-Hill, New York, 1983.

Peck, R. B., Hanson, W. E., and Thornburn, T. H., *Foundation Engineering*, John Wiley & Sons, New York, 1974.

Powers, J. P., *Construction Dewatering*, John Wiley & Sons, New York, 1979.

Roister, D. L., *Landslide Remedial Measures*, Tennessee Department of Transportation, Nashville, Tenn., 1982.

Sacks, A. M., "Geosynthetics," p. 14, *Remodeling Magazine*, Hanley Woods Inc., Washington, D.C., November 1987.

Sacks, A., "R_x for Basement Water Problems," *The Family Handyman*, St. Paul, Minn., pp. 36–40, September 1981.

Schuster, R. L. and Kruse, R. J., eds. *Landslides Analysis and Control, Transportation Research Board*, Special Report 176, Washington, D.C., 1978.

Schwab, G. O., Revert, R. K., et al., *Soil and Water Conservation Engineering*, 3rd Edition, John Wiley & Sons, New York, 1981.

Sowers, G. F., *Introductory Soil Mechanics and Foundations: Geotechnical Engineering*, Macmillan Publishing Co., New York, 1979.

Todd, D. K., *Ground Water Hydrology*, John Wiley & Sons, New York, 1980.

U.S. Department of Agriculture, "Soil Conservation Service, Drainage," Chapter 14 in: *Engineering Field Manual*, USDA, Washington, D.C., 1969.

USDA, *Soil Conservation Service, Technical Guide*, Section IV, Standard 606, Subsurface Drainage, May 1988.

U.S. Department of the Interior, Bureau of Reclamation, Ground Water Manual, Washington, D.C., 1995.

Vetch, J. O. and Humphrey, C. Rs, *Water and Water Use Terminology*, Thomas Printing & Publishing Co., New York, 1966.

INDEX